The International Fugitive

To those innocents on the run for their lives, and to those who protect them at great peril to themselves.

Kenn Abaygo

The International Fugitive

Secrets of Clandestine Travel Overseas

Paladin Press · Boulder, Colorado

Also by Kenn Abaygo:

Fugitive: How to Run, Hide, and Survive

Advanced Fugitive: Running, Hiding, Surviving and Thriving Forever

The International Fugitive: Secrets of Clandestine Travel Overseas
by Kenn Abaygo

ISBN 1-58160-035-6
Printed in the United States of America

Published by Paladin Press, a division of
Paladin Enterprises, Inc., P.O. Box 1307,
Boulder, Colorado 80306, USA.
(303) 443-7250

Direct inquiries and/or orders to the above address.

Visit our Web site at: www.paladin-press.com

Contents

Introduction

> "Now I unclasp a secret book, and to your quick-conceiving discontents I'll read you matter deep and dangerous."
>
> William Shakespeare, *King Henry the Fourth*

Gazi Ibrahim Abu Mezer and Lafi Khalil, two terrorists from the Middle East who were caught "less than a day away," according to James Kallstrom, head of the New York FBI office, from blowing up a subway station in New York City in July of 1997, knew all about international travel, hideouts, and border crossings. They knew so much, in fact, that they didn't bother to enter the United States covertly on their mission to kill hundreds of unsuspecting people in Brooklyn but rather did so through the same gates that tens of thousands of people use daily.

Abu Mezer and Khalil filled out immigration forms at Customs, and one of them even admitted that he was a suspected terrorist in Israel and had spent time in custody over there. The other, intelligence reports say, was a well-known suspected terrorist, and both were sympathetic to Hamas, one of the most deadly terrorist groups in the world, known for carrying out deadly suicide bombings.

Did anyone—Customs, Immigration, or the FBI—say or do anything to stop these two from entering the country? No.

Upon crossing the border into the United States, the two men rented an apartment in Brooklyn and began building pipe bombs they intended to detonate on the subway. The only reason they were caught was because, fortunately, one of their compatriots squealed on them the day before the attack was supposed to take place.

Surprised that these two guys could get into the country with such ease? Don't be. But how is this possible? Because of the unwieldy size of the government's intelligence and law enforcement network and the outright complacency, stupidity, and incompetence of many federal employees who serve as INS (Immigration and Naturalization Service) and Customs agents. Even a small-time hood turned infamous murderer can apparently carry out a major political assassination in broad daylight in downtown Memphis, flee the city, escape the country via Canada, and find his way to the United Kingdom, then hide out for two months before being taken in by the authorities. James Earl Ray, who fatally shot Dr. Martin Luther King Jr. as the civil rights leader stood on a motel balcony in Memphis, did all of this and very nearly got away with it. But did he really pull off the hit and a border-crossing, ocean-hopping evasion by himself? Unlikely. *Very* unlikely.

We can use these two examples to show us ways of traveling, hiding, and crossing borders, but we must never think that it will always be that easy. In reality, it is often much more difficult.

In my first two books in this series, *Fugitive: How to Run, Hide, and Survive*, and *Advanced Fugitive: Running, Hiding, Surviving and Thriving Forever*, I laid the groundwork for living the clandestine life. In this book, we will explore the fascinating subject of covert international travel as well as advanced evasion techniques that will keep you free for as long as you wish.

Yes, your evasion might offer some pretty sights, but your skills will nevertheless be tested, such as here in the jungles of Central America.

Whether you are a government agent, corporate intelligence collector, someone who witnessed something you now wish you hadn't, or simply a traveler who wants to learn how to remain unfettered in your journeys for whatever reason, this unique book will help you.

You'll note that this third book in the *Fugitive* series focuses quite a bit on travel by water. This is for two reasons: water travel is often overlooked or incorrectly dismissed as being too difficult, and the first two books in the series covered travel by land extensively.

But be warned. The life of the fugitive is seldom an easy one, and depending on who is looking for you, the fact is in some cases you will be looking over your shoulder for the rest of your life. Let's be realistic. If a government or a crime organization wants you badly, they will use whatever means available to get to you, making your life a living hell. For instance, I know two guys who work for a certain government agency that has its own extremely well-armed and trained army, navy, and air force. They enter other countries without permission on a regular basis and do things to people that, if the media ever found out, would bring down the current administration in a technicolor spectacle of intrigue and disbelief. If one of these guys is after you, buddy, you need to leave the planet. Really. A mutual friend said to me recently about one of them, "If you wake up at night and he's looking at you through your bedroom window with that big smile of his, you're a dead man, and there's nothing you can do about it."

True enough.

And forget about the government's claim that they no longer engage in assassinations. I have a government document beside me right now that recommends the use of botulinum toxin applied to the pages of a target's book to kill him.

The need for this book—the entire *Fugitive* series, in fact—is clear, but still there are people who don't like me, this series, and especially Paladin Press. During a recent interview on the *Today Show*, co-host Matt Lauer said that my books were written for fugitives from justice. Whereas it is true that the information in these book *could* be used by fugitives from justice, I liken such an inaccurate claim by Mr. Lauer to gasoline being used by an arsonist. Gasoline was created to power combustion engines, but it *can* and *is* used by arsonists. Do we, therefore, call for the elimination of gasoline? Certainly not. I began writing this series as guides for government and military special operations personnel, people running from someone who means them harm (a commonplace occurrence, I assure you), and curiosity seekers (as evidenced by my numerous appearances on television, radio, and in newspapers).

So we see that this is a serious endeavor, as serious as it can get. With that in mind, let's begin with border crossings and go from there.

CHAPTER ONE

On the Border

> "I have a rendezvous with Death,
> At some disputed barricade."
>
> Alan Seeger,
> *I Have a Rendezvous with Death*

The scene appeared like something out of a science fiction thriller. Through my thermal imaging sight I could easily see the greenish-white, ghostlike images of the men on the other side of the border, which was heavily mined and wired as well as covered with interlocking fields of fire from dozens of machine gun emplacements and unseen mortars and artillery pieces. I could see the rifles, grenade launchers, and automatic weapons they were carrying, along with their radios, packs, and other equipment. My partner and I were lying on a slight rise, and we were well concealed with desert camouflage uniforms that had been painted tan. While the paint was still wet we had rolled in the sand so that the sand would stick to the paint and add a natural texture to the uniform.

"You got 'em?" I quietly asked my partner, my elbows digging into the cold sand of the desert night.

"Yeah," came the calm, matter-of-fact reply.

"If you miss you owe me a case of my favorite single malt scotch, and I have expensive tastes."

"The HMFIC is, if you will pardon my blasphamous terminology, about to become a martyr," he said, referring to the military's quaint abbreviation for the Head Mother Fucker In Charge. "And if I nail him?"

"What do you have in mind?" I asked.

"You buy me round-trip air fare to Key West so I can go tarpon fishing."

"Deal. Do 'em."

My partner's rifle erupted with a tremendous boom and my vision was momentarily disrupted by the dust and sand rising up from the ground beneath the muzzle. However, the muzzle blast didn't quite obscure my seeing the man's upper torso literally explode, his body hurling backward as pieces of him flew away from his chest, each piece glowing as if someone had just thrown the insides of a chartreuse Cyalume "light stick" into the air.

The men on the other side of the border instantly began screaming and firing their weapons wildly into the night. Four more exploded—one for each of the next four rounds my partner fired from the .50 caliber Barrett Special Application Scoped Rifle, or SASR—before the fools realized how much trouble they were in and began running for their vehicles.

"That's enough," I told the grinning shooter beside me. "We better save our ammo. That stuff don't come cheap, ya know."

"So much for their little infiltration plan, huh?" he queried.

"I reckon. Maybe next time they'll think twice about coming south. Tarpon, eh?"

"Yeah, tarpon."

And that is the key to crossing a border: think twice. Then think again. Border crossings are one of the most treacherous of undertakings you can contemplate and plan for, and without excellent planning, extreme attention to detail, and perhaps some luck, things on the border can get grim mighty fast. It not only got grim for the poor planners whose demise I just described, but graphic as well. I am sure the survivors have a clear recollection of that night on the Saudi-Kuwait border, and I suspect their desire to cross the border was substantially diminished from there on out.

Make a mistake on the border and you may not have the chance to make any more.

COVERT OR OVERT?

The first decision you must make when faced with a potential border crossing is: Do I *have* to cross the border?

This is not a decision to be made lightly. The fugitive must seriously contemplate all his options and look at each of them objectively and thoroughly without allowing emotion or pressure to get to him. If the decision is made that, yes, you have to go across, the next question is: Do I want to cross covertly or overtly?

Don't jump the gun here, buddy. Just because someone you would really rather not ever see again is looking for you and wants *very badly* to find you doesn't mean that he, she, or even they have connections and power along the border you are about to cross and at the very place you are about to cross. Let's look at this.

Let's say, for instance, that you had a gruesome misunderstanding with Louie "The Drill" Pasculli and he ended up getting dead. You were only defending yourself and in a court of law you would walk away unscathed, but Louie's pals don't see it that way and it is not court of law you are worried about. The money you owed Louie is long gone, and now Louie's business partners want to find you in order to arrange for your paying back of the loan, *plus interest* and a hefty *penalty fee*. It is unlikely that common hoods like Louie and his pals have any pull at the border, and even if they did they would have to grease the skids along the entire border to have any hope of catching you. This means that an overt crossing is probably your best bet.

The Overt Crossing

Overt crossings should be kept simple and direct. There is no need to get slick.

- When the border station is backed up with lots of people, park your car a short walk from the station and hoof it across, saying that you are only going for the day on a brief shopping trip in Tijuana, Juarez, or wherever. Or you might select a small crossing station like St. Stephen across the St. Croix River in Maine, where you can just drive across with no problem and very few questions, not to mention the remote likelihood of Louie's reach ever stretching that far north.
- You must *think* when answering a question. Customs agents are fond of asking what your plans are, how long you intend to stay, what's in your trunk or backpack (fruits, vegetables, and other things that can cause agricultural problems are of great concern to them, as well as knives, guns, drugs of course, and other contraband), and whether or not anyone you don't know asked you to take something—like a package—across the border for them. All of these questions should be accurately answered "No."
- Look the customs agent in the eye, smile like you are just another happy American, answer his questions without looking down or away, tapping your fingers on anything, or doing anything else that might make him think you are nervous, and move along.

But what if whoever is looking for you might very well have some pull along the border? Perhaps you were set up by someone you pissed off and now the DEA thinks you are a bad guy. Maybe you stumbled across a band of Mexican car thieves and now you have to get the hell out of Old Mexico mucho fast but you suspect they have bribed the Mexican police a couple of hundred yards short of the border to stop all cars before they get to the U.S. Customs station. Now what?

A covert crossing is called for.

The Covert Crossing

This is where things start to get a bit dicey. Nevertheless, covert border crossings are conducted hundreds of times daily around the world with no problems at all. On the other hand, hundreds of people are caught daily, too. Depending on who you are and your exact situation, getting caught might prove to be a mere inconvenience or it might be fatal, as the would-be infiltrators my partner and I caught learned all to well. And trouble can take many forms, such as when a U.S. Navy underwater demolition team (UDT) member I know, working to insert guerrillas in North Korea, was very nearly caught along with his team when some dogs in a nearby village began barking as the men made their way ashore from the USS *Begor*. Borders and border areas are rife with unknown dangers.

At one time I lived very close to the Mexican-American border. I was always amazed at the number of illegals trying to cross as well as the number of border patrol agents and the support equipment they had (trucks, helos, ATVs, horses, night vision gear, and so on). Because the United States has erected a border system that is not designed to harm people it doesn't want in, the border is relatively safe to negotiate, as millions of illegals can attest despite all the border patrol agents and their toys and gadgets. But many international borders are designed quite differently, built to maim and kill and therefore seriously negate most attempts at infiltration.

One such border lays between North and South Korea, where I have spent quite a bit of time. One night nearly 20 years

The U.S.–Canada border is best crossed either overtly or via a remote section of woods. This photo was taken approximately 28 miles from the border of Maine and Quebec, where, after crossing the Kennebago Divide, the fugitive could slip over the border south of Dennison Bog.

ago on the beach at Kangnung—a town just south of the DMZ (demilitarized zone) on the eastern side of the peninsula—some North Korean commandos sent to infiltrate (and do what else?) South Korea thought it a good idea to avoid trying to cross the border on land and instead tried to infiltrate via

Comparatively few extreme northern borders have fences and other security systems because the countries who own them rely on Mother Nature to defeat border crossers. However, skilled evaders can make do and get across.

scuba. Their bodies ended up littering the beach in what was a brief but very intense firefight. Years later a North Korean submarine filled with commandos ran aground in the same area and all but one, I believe, were either killed by their own leaders (presumably to keep them from talking; I have witnessed interrogations performed by South Korean professionals and hope to never be on the receiving end of one of those conversations) or by South Korean security forces.

The fugitive must absolutely avoid attempting to cross such heavily defended and patrolled borders covertly, that is, with the notion that border agents won't know you have entered the country. Such borders, if they must be crossed, should be crossed as routinely as possible right at a regular border crossing. This was precisely how I entered and exited East Germany from West Germany during the Cold War, and I seldom had a problem. And when I did, pleading ignorance and not raising a ruckus always worked, such as the time I was seen taking photos of this and that in East Berlin. When I crossed back over they took my film *and* my camera and lenses, which I had hoped would be seen as just the trappings of another tourist. They weren't, but I was allowed to return to the safety of West Berlin otherwise unmolested.

Covert border crossings are exceedingly dangerous, and if you are reading this latest book in my *Fugitive* series, then there is a chance someone wants you and that someone is someone you never want to succeed in his quest. This means that you are going to have to go to considerable lengths to stay free, so in case you *are* desperate enough to attempt a covert crossing, let's take a close look at all the factors you must consider first.

Patrols

I wish I had a dollar for every hour I have spent on patrol. But I don't, so let's skip the wish list and get down to brass tacks.

A patrol consists of anywhere between one and usually no more than a dozen men (a squad) if they are on foot, with one- or two-man patrols usually being conducted by vehicle. Their job on a border is to catch people like you or make people like you decide not to even try it, and there are innumerable ways to achieve both ends. First we will look at the foot patrol.

Foot patrols normally consist of:

- a patrol leader
- an assistant patrol leader
- a point man
- "tail end Charlie" (the last man in the patrol, often armed with a grenade launcher)
- a radio operator
- a couple of riflemen (one with a second radio)
- a couple of grenadiers
- a couple of automatic weapons men
- a medical specialist

You'll note that this patrol adds up to 12 men, but there could just as easily be six or 15 or any other number depending on a variety of reasons, which aren't important because there is nothing you can do about the number of men in a patrol. What they show up with, they show up with, and you have to deal with it.

Never assume that a patrol that had 10 men last night will have 10 men tonight. An old trick is to go out with a certain number in your patrol two or three nights in a row and then change the number substantially. This is used to make the fugitive think, after surveilling a border area for a couple or three nights, that once he sees what he thinks is the last man pass his position, it is safe to cross. Little does he know that a final man is trailing his buddies quite a ways back in the hopes of spotting someone like you cutting behind them.

Patrols walk in a certain order dictated by the patrol leader. What follows is an example of a typical order of movement for a patrol, but it in no way means that this is what every patrol will look like. It simply describes how I set up my patrols when I hunted people like you a long time ago.

The point man will likely be carrying a rifle and will certainly be the most stealthy man in the patrol, with the best eyes, best attention to detail, best ears, best nose, and best sense of something being not quite right, which is the sense most fugitives fall prey to. It is that inate feeling that someone is nearby or someone has passed this way or someone is watching you, and it is a real killer. I am told by anthropologists that this feeling is a throwback to the days when we humans were a bit more hairy and skulked about trying to beat critters to death with sticks. It was a time when we were preyed upon by other mammals and had to rely upon this inate survival sense to live to see the sun come up. Good point men still have this sense, and it is very sharp.

The second man is an automatic weapons man carrying a light machine gun. I always liked putting one of these guys close to the point so that I could get a lot of rounds down range quickly should the point man need them.

The third man is the patrol leader. Yes, this far forward in the patrol. By being here the patrol leader is close to the point man, making directions and orders easier to relay. He also has a better sense of what is happening up front.

Behind the patrol leader is the primary radio man. It is often a necessary evil to have the primary radio man this close to the patrol leader so that he can communicate with minimal fuss and muss. The danger is, of course, that an enemy seeing the radio man always going to a certain man in the patrol will know that the man always being handed the handset is in charge. That's how we identified the leader that night on the border of Saudi Arabia and Kuwait. He kept walking back to his Panhard scout car to talk on the radio. (He also had binoculars draped around his neck. Anyone with binoculars draped around his neck who is always talking on the radio is worth the price of a bullet, in my book.)

Next comes a grenadier, and he is followed by a rifleman. Moving on down the patrol you have the assistant patrol leader, followed by the medic. Then comes another rifleman (with the second radio), a grenadier, the other automatic weapons man,

The author scans a very remote region while laying out potential evasion routes should things get ugly.

and finally tail end Charlie, whose grenade launcher should be loaded with an antipersonnel round.

That's quite a bit of firepower backed up by 12 sets of eyes, *all looking for you*. Add the fact that they might have night vision devices (NVDs) and thermal imaging sights, then throw in their knowledge of the area they are patroling and the annoying possibility that they are working with other patrols in the area, and you have 12 men that you need to avoid by a long length.

Smart border patrols have many tricks up their sleeves. One is the "drop" sniper. This is when a patrol stops for a rest and leaves behind a sniper team—"drops" them—when they get going again. This can be a highly effective ruse and it is one that you must be alert for. Count the number of men in the patrol as soon as you see them and make sure the same number leaves after their rest.

A variation of the drop sniper is the distant overwatch. Used when the patrol will be in terrain lower than that surrounding the patrol route, an overwatch team is used to watch the patrol as it passes through an area. The patrol proceeds as if nothing is the matter, but the overwatch team, surveilling the patrol route with night vision devices and oftentimes a sniper team, keeps an eye out for someone trying to sneak behind the patrol after it passes a certain point. The overwatch team can then contact the patrol on the radio when someone is spotted and have them circle around to cut him off, send another patrol to the area, or simply have the sniper team shoot the infiltrator.

Night vision devices are one of the most dangerous things the fugitive will have to deal with when crossing a border, and patrols along modern borders will likely have them. The best answer to counter a patrol's NVDs is to have your own so that you can see them before they see you. Also, by traveling during a thunderstorm you can avoid detection by NVDs because most will white-out (become temporarily unusable) when exposed to lightning flashes.

Regular night vision goggles (NVGs) illuminate what the wearer is looking at, but they do not show body heat. This means that a camouflaged and stationary man being viewed by NVGs may not be detected because he appears to be just another shrub or log or what have you. Thermal imaging systems, on the other hand, detect body heat (as well as the heat coming from an engine, fire, or the barrel of a weapon that was just fired). Patrols using thermal imaging systems, therefore, force the fugitive to use terrain to mask his whereabouts and movement. By sticking to arroyos, draws, culverts, and streambeds, or behind knolls, hills, ridges, and so on, the fugitive reduces his chances of being detected by thermals. But here is where the game takes yet another twist.

If you were the patrol leader and you suspected someone was trying to get past you by avoiding your thermal imaging system, what would you do? Naturally you would take extra steps to cover areas masked from your thermals. This would include placing surveillance or sniper teams near low-lying and otherwise obscured approaches or perhaps running extra patrols through the area or covering it with booby traps and antipersonnel mines. It might also include inserting acoustic sensors or using ground surveillance radar (GSR) to cover the area.

GSR is literally radar that detects movement along the ground. It has come a long way since its inception and today, experienced operators with good equipment have no problem picking up a man walking through the brush of west Texas or anywhere along the Mexican-American border that isn't masked in some way. Joint Task Force (JTF) 6, the border surveillance unit of the U.S. military that assisted the U.S. Customs Service and U.S. Border Patrol to stop drug smuggling and illegal aliens, used ground surveillance radar to support foot, vehicle, and air patrols. (All such military assistance to law enforcement on the border has ceased in the wake of a civilian shooting by a U.S. Marine reconnaissance team.)

When you combine GSR, NVDs, thermal imaging systems, and aircraft as well as human intelligence, you get a border that one would think would be unbeatable. Nevertheless, thousands of illegal aliens and astounding quantities of illegal narcotics are smuggled across the border every day, and the number of ways of beating the border—any border—are incredible.

For example, I once crossed a border mixed in with a herd of alpacas, a domesticated animal something like a llama. GSR was being used on that border, but to the radar I was just another alpaca. When they came near the fence I dropped to my belly and slithered under in a dry wash that I had seen in an aerial photograph of the area. I crawled along that wash for several hundred yards before coming to a stream, which I floated down to its confluence with a mid-sized river, which was my haul-out point. From there I had only two clicks (2,000 meters) to reach my planned lay-up point from where I could complete my target folder. I would have had to try something else to infiltrate the border had those alpacas been wild vicunas or guanacos, but luck was on my side that night, and luck counts as much as skill.

How did I know that no patrols were in the area? Well, I didn't, exactly, but the same aircraft that took the standard aerial photo took a thermal image, which showed no human activity except for at the target itself and the muddy two-track road that led up to it from the outskirts of the town of Puno.

But any patrol's biggest weakness is the fact that the members of the patrol are human. This means they have certain predispositions, some of which you may be able to exploit.

Take the weather. A miserable patrol out in a storm is a patrol more likely to be walking with heads down trying to stay dry and warm. This means that your best time to move is a night with driving rain, wind, and cold. Heavy rain and wind also helps wash away your tracks and other evidence of your passing, and inclement weather can help minimize the usefulness of NVDs. But you must be careful of flash floods, as several illegal aliens who had just crossed the border in Texas found out recently when they were killed in an arroyo that flooded suddenly when they were in it.

Fences

I was once an instructor who taught people how to get through fences, among other things. I always started off by telling my students that the reason the fence is there is because someone either doesn't want you getting in or doesn't want someone else getting out. Sometimes it is both. In the case of

East Germany, the fences were for keeping people—the East Germans—in. In the case of North Korea and the old Soviet Union, the fences were for keeping their own citizens in as well as keeping people—those bad NATO hooligans—out.

It is important to know why the fence was erected so that you can better formulate your plan. You see, a fence designed to keep people in is different from one designed to keep people out. This is the same principle that vault doors are designed with; it might be very hard to open from the outside but preposterously easy to open from the inside. Remember, too, that you could be trying to either get out of or into a country.

The best fence system is the one that doesn't look all that hard to get across or through but which in reality is more dangerous than it appears. The last fence I had to cross under duress was in fact a wall. It was made of cinder blocks and was about eight feet high. When someone started shooting (come to find out later that they weren't even shooting at me; silly me), I ran for it and headed straight for the wall. Dragging my tired carcass over the top, I found that someone had set broken soda bottles jagged-end up into the top of the wall and had set them in with cement. I haven't a clue as to how I didn't cut myself up there, but if luck is with you, then keep going as fast as you can until it runs out. I dropped back down on the other side and kept running.

Now, if I had been the guy who set those broken bottles into the top of that wall, I would also have set antipersonnel mines all along the base of it. That way if someone did get over the wall, they would get a funny surprise when they dropped over the other side. The wall I scaled that night in the Middle East reminded me of the one around Shanghai's Qing Pu Prison (Prison Number 5), only not as nasty (Qing Pu's glass shards are much larger).

Fences can also appear to be simple to negotiate, but in reality they are treacherous. Such a fence is the Innofence, manufactured by an Israeli company called Magal Security Systems.

I have run into these things twice: once in the Negev Desert near Mizpe Ramon (a short jaunt from the Sinai border) and the other time near Teverya on the Israel-Syria border. I could have just hopped over both of them but didn't right away because I recognized them for what they were: fences with built-in fiber optics that tell a security guard somewhere nearby that someone just went over, through, or under the fence and precisely where they did so. Still, I had to get to what was behind both fences, so I reconned each area to find the guards' command center. Then I surmised where the fiber-optic cable that linked the fences to the center was run underground, and I dug a narrow (four inches or so) trench perpendicular to where I assumed the cable was until I hit it. I cut the wire in each case. Naturally, there is a built-in alarm system that tells the guards something is wrong with the system, and naturally they come to check the fence. But by that time I had filled the hole back in, erased my tracks, and scampered to cover until the guards went back to their hootch. I did what I had to do and left.

Another slick trick that can be used to modify existing fences is a set of wires that are connected to piezo-electric strain detectors. When strain is placed on the wires, a signal is sent to a microprocessing unit and is relayed to the guard shack. What to do? Fool the guards. Although the wires are set to ignore things like foul weather and birds that can effect the system, human nature can still be exploited by making the guards *think* that the weather or some birds or squirrels set the system off. You can drop a limb from an overhanging tree onto the wire after you get through, if, that is, there are trees overhanging the fence, which there really shouldn't be. The guards will check on the disturbance, find the offending limb, and reset the system. By that time you will have rubbed out your tracks and moved on your way. Use your imagination.

Sometimes it can appear that no fence is even there, when in reality it is simply invisible. Such fences are usually made of laser beams fired from a transmitter to a receiver. If anything interrupts the beams, the system alerts the guards to the fact that something is going on and tells them the sector it is going on in. You really have to watch these systems, as they can be hard to beat. Do a thorough recon of the site. Look for posts that might house transmitters and receivers. Once you think you are confident you know their locations, use handfuls of baby powder blown into the air near where you think the beams are to detect them, then find a way through. This method was dramatically demonstrated in the opening scene of the great Peter Sellers movie *Return of the Pink Panther*, except the cat burglar used an aerosol spray can to illuminate the beams.

Infrared systems are also in use at borders. They can be more dangerous to the fugitive than laser systems because the unaided human eye cannot detect infrared light. Infrared systems have been in use since World War II, and they work, too. They function on the principle of heat variance detection, i.e., they detect and warn of a sudden, localized increase in temperature, such as body heat or the heat of a mechanical device of some kind, like a vehicle, boat, or aircraft. (This is the technology used in military aircraft FLIR systems, FLIR standing for forward looking infrared.) Different systems have different focused zones, with some being 360 degrees and others covering areas in the shape of cones or rectangles.

To beat such a system, you must first recon the area and find where the pylons are located (some systems use a single pylon). These pylons are usually a few inches thick and are about three feet high, and they might be camouflaged. The best way to circumvent the system is to use a Mylar blanket as a screen, keeping it between you and the sensor at all times. The Mylar blocks the body heat you give off and therefore prevents it from being detected by the sensor.

Warning: It is common practice to use three or four seemingly redundant systems at one time, such as infrared and laser systems backed up by ground surveillance radar and perhaps acoustic sensors and things like pressure plates in the ground.

Fences designed to keep others in will have a cleared area leading up to them so that anyone trying to make a run for it can be more easily seen and fired upon. Countries that have such fences frequently erect false borders that look like genuine border fences but which are in fact hundreds of meters shy of the actual border. Why do such a thing? To give the fugitive a false sense of security and encourage him to let his guard down. The area between the false border and the real border will be heavily mined, of course. At the end of the cleared area will be a series of obstacles that will get progressively more difficult to negotiate, such as tanglefoot (wire strung in individual strands between 10 and 26 inches high designed to trip you up and slow you down), standard barbed wire, and concertina (razor wire), as well as ditches that are just a bit too wide to jump across filled with murky water in which has been placed more tanglefoot, concertina, sharpened stakes, antipersonnel mines, broken glass, and the like. Walls of tight barbed wire set about three and a half feet apart (which is longer than the average human arm) will likely come next, followed by another cleared and mined zone and finally some more wire walls and concertina. As a real-world example, Israel's Beer Shiva Prison has multiple fences of different sorts with dead space in between and a final obstacle in the shape of a tall wall. If you get out—and that isn't all that likely—Beer Shiva's notorious dogs will be on you like a Little Rock lawyer on a bogus land deal.

One border I crossed even had a 15-foot-deep ditch with poured concrete walls that was about 15 feet across. It was dug only a few feet from the last wire wall so you couldn't get a running start to even attempt to jump over it. How did I get over it? I tied one of my pistols to some parachute (550) cord and threw the pistol into a tree on the other side. I then tied a quick bowline around my chest with the other end of the cord and dropped down into the ditch after taking up the slack. Then I just climbed up the cord using "quick-step" knots (which take about one second each to rig), crawled out of the ditch, and ran for the tree line. I was fortunate that no one was chasing me at the moment, which would have made things much more exciting. Of course, had I been better prepared and had better intel, I wouldn't have been throwing one of my pistols around with a piece of rope attached to it, now would I?

Barbed wire—the same stuff used on those old fences to keep cows in the pasture—is more difficult to employ than concertina wire, but when a barbed-wire fence is erected properly, the obstacle can be maddeningly difficult to negotiate. The double-apron barbed-wire fence is quite common along borders. It consists of a front apron, center fence, and rear apron, all about four feet high, and measures about nine feet from the edge of the front apron to the edge of the rear apron.

A double-apron fence can be a real pain, but it is negotiable. The end of the fence is usually the most easily thwarted section because here all the wires come together and are fastened to an anchor picket. A fugitive's best bet is to cut through the wires a few inches before they attach to the anchor picket, or, if the men who built the fence didn't do a great job anchoring the picket into the ground, the fugitive can wiggle it back and forth and pull it out of the ground.

If you happen to have something that will suffice such as a wide board, you can throw it over the wire and scurry over the top. The Vietcong often used heavy blankets and such things to hastily negotiate double-apron and other barbed-wire fences when in the attack. Sometimes a guerrilla would even run up to the fence and throw himself on it to allow other guerrillas to run right over his back to assault and breach a perimeter.

Concertina fences are also commonplace and are a favorite of prisons, including the "SuperMax" federal penitentiary near Canon City, Colorado, which I am familiar with. Concertina is bad stuff—much more bothersome than a double-apron. It is used so much because it can be hard to get through and is easy to deploy, since it comes in rolls that just roll out, requiring little effort or know-how to set up.

There are two primary types of concertina: standard and razor. Standard has typical barbs like those found on your usual barbed wire; razor wire has insidious little razor blades attached all over it, which can really degrade one's karma in a big hurry. The latter is the shiny rolls of wire you see piled up in stacks around the perimeter of prisons and on borders everywhere, it seems.

But even razor wire is beatable if you have the time and a few tools of the trade. For wire that is staked down, spread the strands and place a long stick with a notch in each end in between them to hold them apart, then slide through the spread section and continue the process. You can also use cord to tie the strands apart after spreading them. For wire that isn't staked down, a notched stick can be used to pry up the wire and hold it up while you slither underneath.

A wire blanket is another option when dealing with fences of all sorts. It is nothing more than a heavily padded blanket that you drape over the wire and then crawl over. The weave of the blanket must be very tight to prevent it from being cut or snagged by the razors or barbs. Make sure you check the wire after you pull the blanket off to see if any fibers or pieces of cloth were left behind, which will alert a sentry to your passing. As with all barbed wire, when fiddling with it for any reason, wear gloves.

Any fence can have a track trap (soft sand on either side of the fence), so watch your step. You must be prepared to eliminate your tracks whenever operating near a fence, and the time to think about how you are going to wipe out your tracks isn't when you are halfway over the fence. A large soft towel is excellent for this purpose. Have one with you and ready for immediate use.

Pay attention to signs, too, and hope you can read what they say. For instance, the skull and crossbones symbol is known everywhere as a danger signal, but would you know what accompanying Arabic writing might say? Well, it might not say "camel crossing." It might say "mines."

Culverts and Ditches

Culverts and ditches that run under borders are seldom easy to negotiate, as they tend to be fenced, booby-trapped, or mined. The fugitive must be prepared to deal with any eventuality.

Culverts often have water in them, but they may be seasonal insofar as the amount of water goes, with more water during the rainy season or, in mountainous areas, during the run-off (snowmelt). Also, in desert regions flash floods can fill an empty culvert in seconds and wash you away like so much flotsam.

You have to be ready to cut wire and bypass or otherwise thwart booby traps and mines when using a culvert. As they are weak points in a border security system, you can bet that patrols or electronic systems will be monitoring each one. This means that the best time to use one is when the security systems are least likely to expect one to be used, such as during a bad storm. Yes, you are going to be wet and cold and miserable, but it is for these very reasons that you want to use the culvert when its nasty outside; the patrols will be less inclined to check the culvert thoroughly, and cameras and sensors will likely be substantially less effective because of the rain, wind, hopefully thunder and lightning, and reduced visibility.

Warning: water in culverts is notoriously fast-moving and treachorous. If you are swept downstream, flip onto your back, and keep your feet downstream and your arms perpendicular to your upper torso. Use your feet to shove off rocks and other obstructions and your arms to maneuver. Watch for eddies to swim into to get out of the water. You want to at all costs avoid getting pinned against a grate in front of a drainage pipe or entangled in debris. If this happens, you *will* drown.

Ditches are easier for security patrols and monitoring devices to check since they are open on top. All the same rules apply.

Culverts and ditches are natural pathways for animals. It might help to watch one for a while to see how coyotes, deer, and other mammals are using them as corridors. And remember that culverts and ditches are natural collectors of objects washed downstream, such as broken bottles and other debris. I recall using a culvert on an island in the Pacific in 1979 when someone I was with stepped on a broken bottle. It cut right through his sneakers and badly cut his foot.

Rivers and Streams

On that very same island I once had to ford a small coastal stream running next to a road. There was some light brush on the other side of the road and we were trying to make it across the stream and road as quickly as possible in order to get to that brush. We were in a big hurry and didn't put out security to cross the road, which is a major mistake. We made it across and dashed for the brush, but just as we all stood up to run across the road, a truck came around the blind corner. We instantly crouched and pretended to be rocks beside the brush as the truck passed. It worked, although I have no idea how the men in the cab didn't see us. I fully expected to be shot right there.

The lesson: don't let the pressure of time constraints allow you to make stupid mistakes.

It is very easy to booby-trap a river or stream. A strip of razor wire run across a stream just under the surface can cut you to shreds. A trip wire rigged to an explosive charge is simple to set up. There are more ways to rig a booby trap across a stream than can be listed here, so you must go slowly and use the utmost caution. It might look easy, but the chances are that it isn't.

In rivers and streams with current, use a log with branches below it to travel downstream. In fact, use two. The first one is sent downstream ahead of you to hit any booby traps. The second is used as flotation and camouflage for you. Stay at least 100 yards away from the log in front of you to avoid shrapnel from any exploding booby traps. And don't kick with a flutter kick to propel yourself, since kicking in this manner makes splashes and draws attention from people or things in the water or on the bank that might want to eat you, like crocs and gators, if those happen to be indigenous to the area you are in. You can disguise yourself on a log with brush attached and float right by such critters without their coming to investigate. If you must propel yourself, use a breaststroke kick or smooth, steady scissors kick.

The chances are slim that you will find yourself using scuba gear to cross a border, but it is possible. Get training before you ever use scuba, since laws of physics like Boyle's Law, Charles' Law, and Dalton's Law can all kill you if you don't know what they are and how to get by them. Furthermore, if you know what's good for you, you will always use what's known as closed-circuit scuba, which leaves no bubbles to float to the surface. But as I said, you will probably never have to use this gear.

Boats and other watercraft might be an option to foil a water border. Stealing a boat is often very simple, but you have to take all the standard precautions. First, what size boat do you need? Use the smallest possible to attract less attention. Check the gas first; make sure there is enough. Pick one that doesn't require a key (a small outboard motor on a skiff does not require a key) or, if you know how it's done, hot-wire one. Canoes and other self-propelled designs can be just the ticket as well.

When using a boat, as always, consider who will be watching for you. For example, if I needed to get back into the U.S. from Mexico, I probably would get to Tijuana and steal a boat. I would then travel north until off Point Loma and turn in toward San Diego Bay, then go straight toward one of the docks used by restaurants lining the bay. I would tie up and walk away as if nothing were the matter. By heading over the horizon first, you bypass the prying eyes watching from the beach at San Ysidro and Imperial Beach.

A word of caution when trying this stunt. As you enter San Diego Bay, on the right is North Island Naval Air Station, which has plenty of security that you have to avoid, so don't go ashore there. On the left are submarine pens, which are obviously demanding of substantial security. Also on the left are dolphin (porpoise) pens containing highly trained mammals that belong to the U.S. Navy. These creatures are top secret just like the subs near them, and security is tight. Stay clear of them. I'd tell you what these dolphins are capable of, but if I did some government types would eventually track me down and have words with me, then I would go away with them and lose out on all my royalties. We can't have that, now can we?

Trains

If you are old like me, then you are familiar with the term *hobo*. A genuine hobo is a bum who travels a lot by train, and in the old days one could easily ride the rails all over North America. Today things have changed, especially when it comes to hopping a freight across the U.S.-Mexico border, where Border Patrol agents check every train car. Other borders, on the other hand, can be quite easy to cross on a train that you technically don't have permission to be on. Europe is a good continent to try crossing borders on trains because so many people—locals and tourists alike—do so every day.

Forget about hanging underneath the car. That's a great way to get killed needlessly. Instead, get in the car and hide among whatever is in there. Most cars will be locked in some way, so the thing to do is toss out some of the merchandise that is in the car at night in order to make it look like the car was broken in to for theft purposes. Then leave the door slightly ajar as if the thieves didn't bother to close it all the way. The idea is to make the searcher think that someone may have ripped something off but isn't there any longer. If the merchandise is in large cardboard boxes, toss whatever is in one of them out and get in. Before closing the box, pull another partly on top of it and then use your handy bottle of super glue to reseal the lid on your box. You will still be able to easily force your way out when it's time.

Planes are an option that require special consideration. Remote airstrips, such as this one in Costa Rica, can be useful, however.

Aircraft

Tricky. If you know how to fly, it is a simple matter to go to a small airport with no tower and hot-wire a plane. If you have to fly over a border, it gets trickier. I generally advise against this.

The dubious trick of hiding in the wheel well of a jet is stupid, since temperatures at the altitudes jets fly are extremely low. You will almost certainly die from hypothermia if you are not crushed by the gear itself.

Some borders with mountains near them can be crossed by hang glider or ram-air parachute. You must know what type of radar is in use before you try this ploy, and you have to have a take-off point and a safe landing point. Reconnaissance is vital. You also, naturally, have to be trained in the operation of a hang glider or ram-air chute, as one mistake can mean a fatal crash.

Cars, Trucks, and Busses

Hiding inside a car, truck, or bus can be very tough to pull off; it just depends on the border you are crossing. Border guards have seen every trick in the book, so you have to really weigh the risks. Because the guards have dealt with crafty drug smugglers and illegal alien "mules" for years, it is very unlikely you will be able to devise something they haven't already seen many times before. Afterall, unless you embezzled millions of dollars from Merrill Lynch, it is unlikely you will have the same resources and cash as the old smuggling pros. Your time and money would be better spent acquiring a bogus passport and crossing right under the guards' noses.

Animal Sentries

Animal sentries can be the toughest of all to get by on a border. Dogs and especially horses can smell or hear a human coming much easier than a man can. You must reconnoiter the border first to identify and avoid these animals. Guards on horses know their animals and know when their horse smells something funny.

One of the best animal sentries is the guinea fowl. These very alert birds scream like mad when they see someone they don't know. Their owner is going to know when they are upset and come to have a look.

CHAPTER TWO

North American Travel

> "To earn you Freedom, the seven-pillared worthy house,
> that your eyes might be shining for me . . ."
>
> T.E. Lawrence, *Seven Pillars of Wisdom*

I have spent the last few days appearing on various news shows, including *Today*, *Extra!*, *Cochran & Company*, *AM Live*, and on CNN as well as local television channels, and have been interviewed by the *Miami Herald* and other newspapers, all because I was apparently the only expert to predict in the national media how spree murderer Andrew Phillip Cunanan would attempt to remain on the lam and escape the dragnet set for him in south Florida. Beginning with a *Miami Herald* interview on July 21, I predicted that Cunanan would remain in the area and attempt to simply hide until the furor died down somewhat, and then suggested that he could make his way to the Atlantic Intracoastal Waterway, board a boat, and never be heard from again. He did just that, only the reason he was never heard from again was because he committed suicide on that boat.

It seemed that everyone

wanted to know how Cunanan could remain a fugitive from the FBI (he was, of course, at the top of the FBI's 10 Most Wanted List) and thousands of other cops in south Florida, not be seen on the street or anywhere else by people who had undoubtedly seen his face on wanted posters everywhere, get food and water and money, and so on. And it seemed that everyone wanted to know, after my prediction and listing of Cunanan's options, how I was able to be so accurate.

Well, let's look at the situation:

- Cunanan was an urbanite who enjoyed the soft, cushy life and all the baubles and bangles it brings; he had no training in outdoor skills. So, we can surmise he would stay urban because of his background and the fact that he would be very intimidated in the Everglades, 10,000 Islands area, or in the undeveloped savanna of south-central Florida.
- We also know that the dragnet went up for him almost immediately after he gunned down fashion designer Gianni Versace on the front steps of his home in Miami's trendy South Beach area, so it was unlikely he would be able to leave the area in a car.
- Miami Beach is located on a barrier island, meaning that to leave by car he would have to cross one of the bridges connecting it to the Florida mainland. Such chokepoints are dangerous, and he probably knew it.
- His face was on wanted posters everywhere, so everyone knew what he looked like and was watching for him. He therefore had to get out of sight for a while or radically change his appearance. He chose the former.
- He didn't want to risk (and probably wouldn't have been comfortable) masquerading as a homeless person and using shelters and soup kitchens, so he chose a place to hide that had food, water, and a telephone

(which he used to call a friend whom he hoped would assist him in acquiring a fake passport): the houseboat.
- The houseboat was on the Atlantic Intracoastal Waterway, giving him easy access to other boats that he could steal and then head north or south in at night. The police were not watching the waterway like they were the highways.

Given these facts, I was able to come up with a plausible evasion plan of action for Cunanan and called it right on the money. No magic; just deduction, a sound evaluation of Cunanan's skills and personality as well as the situation, and a bit of luck.

I was asked by a few interviewers what mistakes Cunanan had made. It was obvious that he wanted the FBI and cops to know he killed Versace. He left his thumb print, full name, and current address (at the Normandy Hotel) at a pawnshop where he pawned a gold coin stolen from his Chicago victim. This was his first mistake: arrogance and a burning desire to be given the recognition he craved. His second mistake was committing a crime not only on a peninsula (Florida) that is easily monitored and cordoned off but on a barrier island; his back was to a wall. Third, he did not take immediate action that first night to steal a boat and blow town by heading out of Miami on the Intracoastal Waterway, which runs from Key West to Pennsylvania. I have spent inestimable hours on that body of water and know that there are tens of thousands of boats of all makes and models on it that are ripe for theft. The waterway is easily navigated at night, too. So we see that Cunanan made three major mistakes, which added up to his richly deserved demise at his own hands.

Because the "hot zone" for your capture consists of where you were last known to be and the routes of most likely escape leading away from there, you must leave the area quickly by a means and on a route that the people looking for you are unlikely to suspect. The idea is get moving quickly, in an unexpected manner, and get out of the country. Andrew Cunanan failed to do this and paid with his life.

But remember: most people on the run are caught when traveling, be it from an apartment to the corner store or at a highway rest stop. Movement is the fugitive's most hazardous undertaking, yet it is one that must be faced all the time. The fugitive isn't going to dig a hole in the ground, jump in, and never come out. He must become a master of anonymous travel.

ANCHORS AWAY

One of the very best means of traveling anonymously in North America is on the water. North America is perfect for such a ruse because of the many waterways available, and most people whose job it is to hunt for you will probably not be looking for you in a boat on Lake Powell, in the Great Lakes, on the West Coast, in the 10,000 Islands region of southern Florida, on the Mississippi or Missouri Rivers, among Maine's or maritime Canada's thousands of islands, or on the Atlantic Intracoastal Waterway, unless, of course, you give them reason to or your name is Captain Ahab, in which case the water is the *first* place they'll look for you. Seriously, if you were a former riverboat captain, ferryman, or commercial fisherman, a boat is the *last* avenue of escape you should consider.

Evasion on the water means you must acquire and maintain your seamanship skills. This includes:

- reading nautical charts
- understanding buoys and other markers
- having boat-handling skills (including in tall water)
- knowing how to maintain your boat (electrical, mechanical, and structural skills)
- predicting weather
- recognizing and avoiding other natural hazards

I have spent most of my life on, in, or around the water and boats of all kinds, and I can promise you that they offer the fugitive a potentially fabulous evasion plan of action.

Houseboats

I once hid out for a week on Florida's St. Johns River on a very large, new, nicely equipped houseboat with a Navy SEAL friend of mine. This was my first time on such a boat, and I was most impressed with the many amenities it offered and struck by the craft's evasion potential. With a little effort, ingenuity, and attention to detail, a houseboat just might be the perfect evasion vessel. But it must nevertheless be rigged for your quick escape should you be boarded.

One means of doing this is to attach scuba gear to the bottom of the boat and use it to make your escape. For you nondivers, yes, it is possible and actually fairly easy to dive into the water and make your way to some scuba gear underwater, turn on the air, stuff the regulator in your mouth, and breath. Once this is done, you can don your mask and clear it, slip on your fins and weight belt, and be under way in less than a minute. I know this because I spent years in special operations in a certain armed force and was taught these skills at a military training center. With training, a man can dive into the water and swim about without a breath for a few minutes, so getting to some nearby scuba gear really isn't a problem. In fact, some friends and I have freedived to get to some scuba gear 80 feet down in the western Pacific.

The houseboat you want isn't too big and isn't too small, meaning that it isn't so big that it is easily noticed and it isn't so small that it won't serve your purposes. It isn't too new and it isn't too old, i.e., a brand new boat is more readily remembered than an older one, and one that is too old will likely have more mechanical problems. It can be easily painted and the name changed, if it has one. Your houseboat should have a draft shallow enough to allow you to get into backwaters and sloughs off the main channel and into shallow water in the lee of an island or a tucked-away cove.

A houseboat on an American river; often an excellent evasion vessel.

The most recent edition of many computer software programs can produce a bogus boat registration with only a little imagination and attention to detail. Foreign registrations are best because someone checking your papers won't be able to access that nation's records. A small country like Guatamala, Belize, or Barbados is best, but you will need supporting documentation to show why your boat is registered there, such as business cards that show you own a company there.

Oceangoing Vessels

That SEAL friend of mine now lives in the Caribbean. Soon he will have saved enough money to buy his dream boat, and he intends to sail off into the sunset and enjoy the good life until he drops dead of overexposure to fun and excitement, no doubt surrounded by beautiful women. I can see him now, anchored off St. Kitts, smiling and having a good ol' time with half a dozen bikini-clad babes decorating his boat. I'll see you down there, Tom.

Oceangoing vessels might include large powered fishing boats, commercial fishing boats, sailboats, or what have you. The idea is to get a boat that you can handle and that won't be missed within minutes. Factors like seaworthiness are crucial, too. Consider some of the following vessels:

Center Consoles

Center consoles are boats with a fiberglass or wood control console in the center of the boat that contains gauges, safety devices, the throttle, and the wheel. Coming in lengths up into the 30-foot range, they offer little in long-term livability for the fugitive, but they are fast and maneuverable and do offer escape and evasion possibilities. Many have a shallow draft that makes it a breeze getting in and out of thin water. A 21- to 27-foot center console like a Jones Brothers Cape Fisherman or a Mako powered by 150–200 horsepower Suzuki outboards and rigged with GPS and a marine radio is a good getaway boat, but you can't live on it for long.

Cuddy Cabins

Cuddy cabins also run into the 30-foot range and offer a small cabin forward that can be equipped with a V-berth, marine head, small stove and refrigerator, and storage areas. Because they can be powered with large outboard motors and don't attract much attention, a cuddy cabin is a frontrunner insofar as evasion boats go. They are also easy to maintain and use less fuel than a larger craft with an inboard engine.

A center console.

Cuddies are extremely commonplace and very versatile. With full electronics, the fugitive can do well in one and can really get around without much fuss or muss. "Skinny water," a mariner's term for shallow water, isn't often much of a problem either, comparatively speaking, because such boats draw little more water than a center console of the same size and draft.

Walkarounds

Walkarounds are bigger and more roomy than a cuddy, but

A cuddy cabin.

the basic layout is the same. Equipped with inboard engines, they have fully enclosed cockpits and can be upgraded with flying towers and many other amenities.

The walkaround is the lower end of boats that can be lived on year round. Built for affordable comfort (a very good one fully equipped starts at about $80,000) and long-range runs, they do suck gas but make up for it in style and space. Because the walkaround has inboard engines, you can expect to have more maintenance duties.

Convertibles and other craft in a Mexican harbor.

Convertibles

Convertibles are the boats you see billfish anglers heading offshore in, with spacious cabins, all the amenities and niceties of home, single or twin flying towers, and so on. They are very expensive and hard to hide. If you own your home and decide to sell it, it *might* bring enough to buy a new or used convertible.

These small yachts require serious seamanship skills and, if they are big enough, you may need a captain's license. They are luxurious, yes, but they are also serious gas guzzlers and require plenty of maintenance from bow to stern.

Yachts.

Yachts

For our purposes, let's call a yacht a yacht. You know, those huge, sleek, floating mansions that come with their own crew in many cases. Yeah, they are nice, but they are also attention-getters in the worst way. Avoid buying or borrowing a yacht. Besides, anyone who can afford one and has his stolen might just come looking for the guy who stole it, adding to your problems. And if you're rich enough to buy a yacht, you should be spending your money on good lawyers rather than books like this to get your ass out of trouble!

A sailboat makes her way along the shore of an island in the Lesser Antilles.

Sailboats

Last winter I found myself in southern Baja. One day a magnificent sailboat of about 130 feet made her way into the harbor. She was something else, her hull a deep royal blue and sporting self-reefing sails and a ready crew. Every day during her stay the owner had the crew take her around the harbor just so folks could get an admiring look at her. Anyone and everyone who went anywhere near that harbor during the boat's stay took note of her lines.

Don't think for a moment that marinas have minimal security.

This is something the fugitive needs to avoid. That is, any boat you select, sailboat or otherwise, shouldn't command attention.

But a smaller sailboat is an excellent idea. They come in an array of designs—ketches, yawls, sloops, and so on—and sizes, and once mastered they can be excellent evasion crafts. Learn how to sail and consider a sailboat from 25 to 60 feet. I have spent time on sailboats—from windjammers off the coast of

Maine to beautiful ketches in the Lesser Antilles—and can attest to their versatility, speed, and comfort.

Canoes and Kayaks

Finally, self-propelled crafts such as canoes and kayaks can be great evasion vessels for the fugitive, but you must consider the pros and cons.

Pros include silence, no fuel requirement, and ease of concealment. Cons are the fact that *you* are the power source and there is minimal space for storage. Nevertheless, they might be the perfect way for you to evade your pursuers silently.

Do not travel by canoe or any other vessel by daylight if you don't have to. Learn to navigate at night, thus eluding most prying eyes.

Sinking your canoe or kayak is nearly impossible due to modern materials that defy sinking quite well. Hiding the craft ashore is a better idea. If you do this well, you could be in East Painlane before they realize they have been duped and spend hours and hours of valuable time backtracking to discover where you exited the water.

NAVIGATIONAL CHARTS AND RELATED INFORMATION

A chart is a map made for the sea and, as such, requires skill and attention to detail to be read correctly. There are many seemingly inconsequential bits of information on every chart that are easy for the novice to overlook, but overlooking the wrong one could mean disaster.

I cannot include in this book an entire course in maritime navigation. However, I will give you the basic information needed to acquire and read charts, which are available from the National Oceanic and Atmospheric Administration.

Canoes on the Allagash Wilderness Waterway in northern Maine. The chances are good that no one would have seen them had they been traveling at night.

Chart and Buoy Basics

The first rule of chart usage is to know and understand all the symbols and measurements on the chart. A chart is a two-dimensional representation of a portion of the earth's surface that shows depths, channels, obstructions, piers, buoys, lights and lighthouses, prominent landmarks, reefs, shoals, sandbars, wrecks, and dozens of other features in and along water. Depths

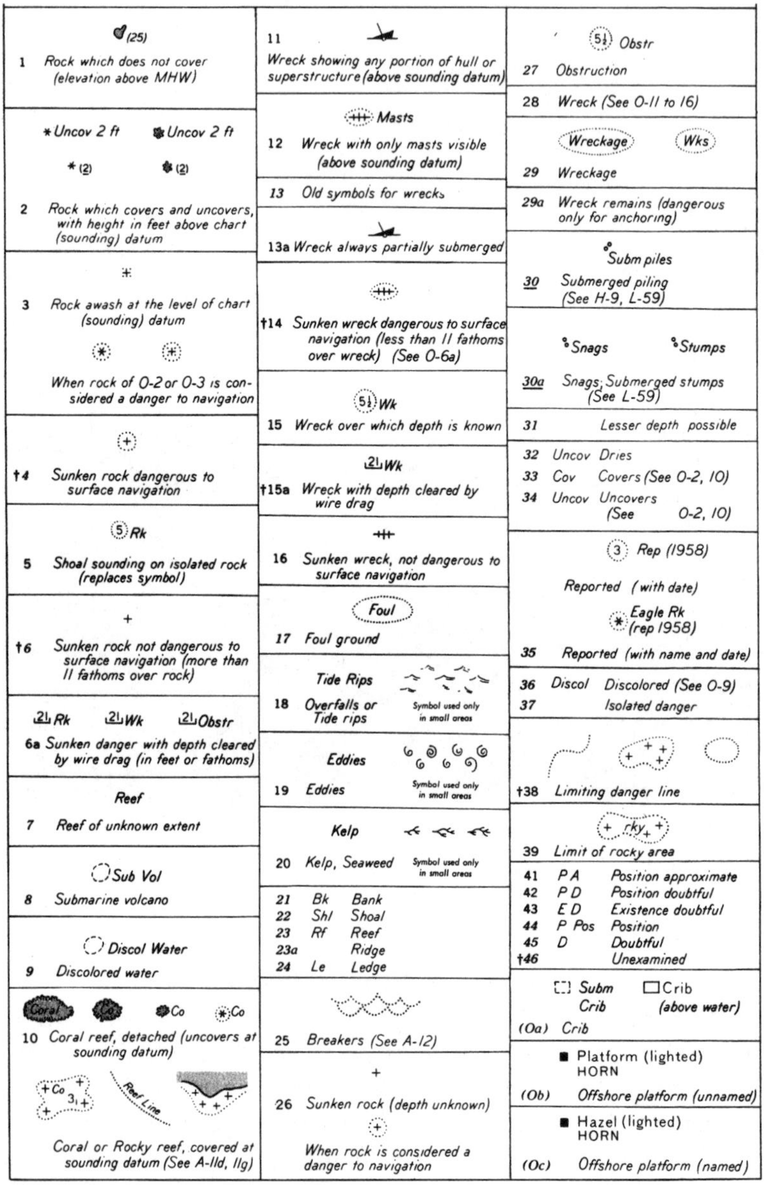

(25)
1 Rock which does not cover (elevation above MHW)

Uncov 2 ft Uncov 2 ft
(2) (2)
2 Rock which covers and uncovers, with height in feet above chart (sounding) datum

3 Rock awash at the level of chart (sounding) datum

When rock of O-2 or O-3 is considered a danger to navigation

†4 Sunken rock dangerous to surface navigation

5 Rk
5 Shoal sounding on isolated rock (replaces symbol)

†6 Sunken rock not dangerous to surface navigation (more than 11 fathoms over rock)

21 Rk 21 Wk 21 Obstr
6a Sunken danger with depth cleared by wire drag (in feet or fathoms)

Reef
7 Reef of unknown extent

Sub Vol
8 Submarine volcano

Discol Water
9 Discolored water

Coral Co Co Co
10 Coral reef, detached (uncovers at sounding datum)

Co 3 Reef Line

Coral or Rocky reef, covered at sounding datum (See A-11d, 11g)

11 Wreck showing any portion of hull or superstructure (above sounding datum)

Masts
12 Wreck with only masts visible (above sounding datum)

13 Old symbols for wrecks

13a Wreck always partially submerged

†14 Sunken wreck dangerous to surface navigation (less than 11 fathoms over wreck) (See O-6a)

5½ Wk
15 Wreck over which depth is known

21 Wk
†15a Wreck with depth cleared by wire drag

16 Sunken wreck, not dangerous to surface navigation

Foul
17 Foul ground

Tide Rips
18 Overfalls or Tide rips — Symbol used only in small areas

Eddies
19 Eddies — Symbol used only in small areas

Kelp
20 Kelp, Seaweed — Symbol used only in small areas

21 Bk Bank
22 Shl Shoal
23 Rf Reef
23a Ridge
24 Le Ledge

25 Breakers (See A-12)

26 Sunken rock (depth unknown)

When rock is considered a danger to navigation

5½ Obstr
27 Obstruction

28 Wreck (See O-11 to 16)

Wreckage Wks
29 Wreckage

29a Wreck remains (dangerous only for anchoring)

Subm piles
30 Submerged piling (See H-9, L-59)

Snags Stumps
30a Snags; Submerged stumps (See L-59)

31 Lesser depth possible

32 Uncov Dries
33 Cov Covers (See O-2, 10)
34 Uncov Uncovers (See O-2, 10)

3 Rep (1958)

Reported (with date)

Eagle Rk (rep 1958)
35 Reported (with name and date)

36 Discol Discolored (See O-9)
37 Isolated danger

†38 Limiting danger line

rky
39 Limit of rocky area

41 PA Position approximate
42 PD Position doubtful
43 ED Existence doubtful
44 P Pos Position
45 D Doubtful
†46 Unexamined

Subm Crib Crib (above water)
(Oa) Crib

Platform (lighted)
HORN
(Ob) Offshore platform (unnamed)

Hazel (lighted)
HORN
(Oc) Offshore platform (named)

Navigational chart symbols.

are shown in either feet or fathoms (one fathom equals six feet) in many locations on the chart; to determine what measurement is being used on a particular chart, look just below the title (name) of the chart. (It is critical that you know if the depth markings are in feet or fathoms. If the chart says you have three feet of water at low tide, but you think the depth marking is in fathoms, you are about to get a rude and potentially fatal surprise.) The chart also tells you heights of objects above the waterline and a wealth of other information you will find important for navigation purposes.

Buoys are road signs for the sailor. Their precise location is shown on a chart as a small black dot that looks like a period. The type of buoy, its number, and the color of its light or type of sound (if applicable) is registered beside the buoy's symbol. These facts are important because in limited visibility you can go to a buoy and determine where you are by reading its symbols and comparing them to those on your chart. Buoys aren't often out of place, but they can move during hurricanes.

Buoy codes are simple. For instance, a buoy marked with a black diamond, the number 2, and FLG 5 sec means that buoy is the black number 2 channel buoy on the port (left) side when heading into the channel. It has a green light that flashes every five seconds.

Rules

As you might expect, there are rules governing the operation of sea vessels that you must be familiar with. Fail to abide by them and you may run aground or collide with another vessel, and that means that you are going to be dealing with the authorities, such as the Coast Guard, which is something you should avoid. The International Rules of the Road (yes, that's the real name) are those rules, and you can get a copy of them from your local Coast Guard or Coast Guard Auxillairy, the Department of Commerce, or perhaps from your local marina. They may be available on the Internet as well. Another set of

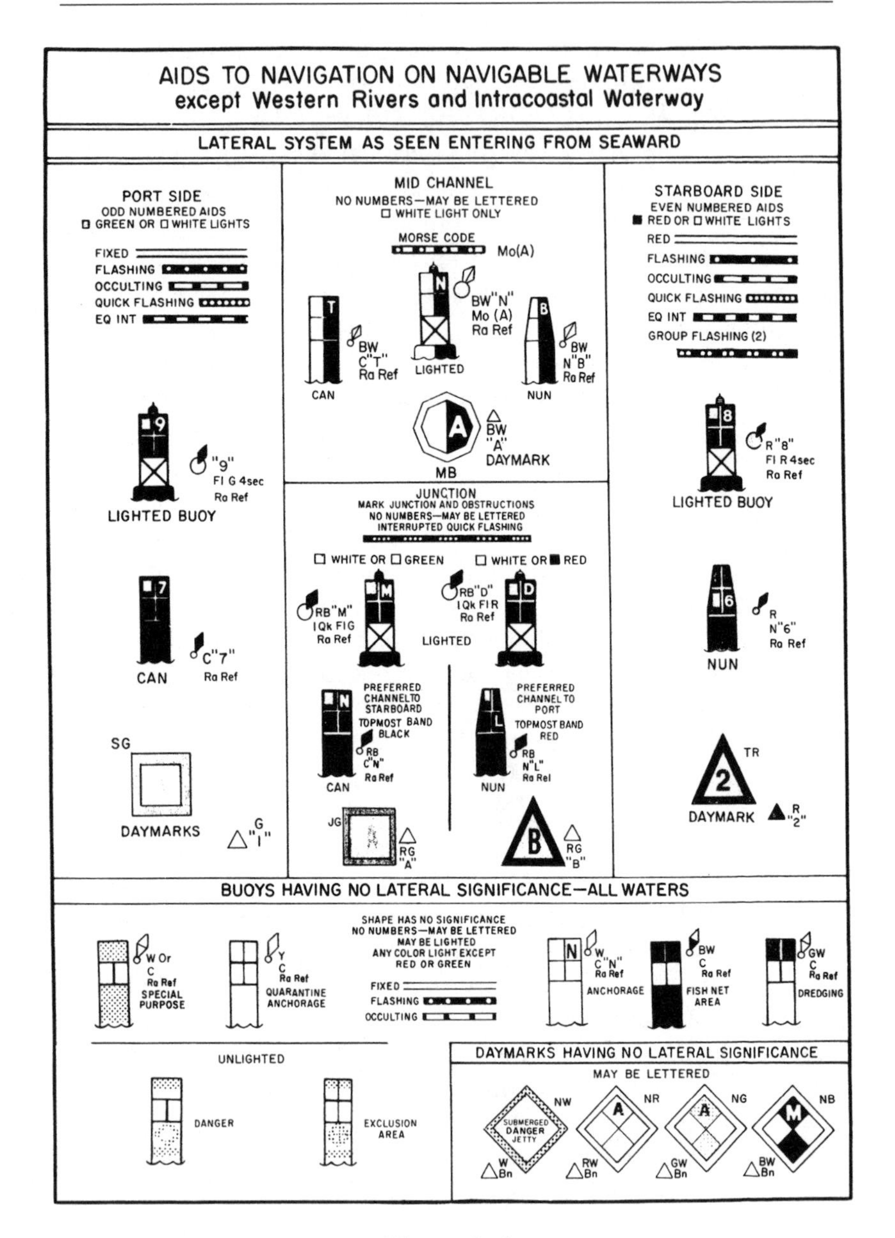

Aids to navigation.

Buoys and Beacons

No.	Symbol	Description
1	•	Position of buoy
2		Light buoy
3	BELL	Bell buoy
3a	GONG	Gong buoy
4	WHIS	Whistle buoy
5	C	Can or Cylindrical buoy
6	N	Nun or Conical buoy
7	SP	Spherical buoy
8	S	Spar buoy
†8a	P	Pillar or Spindle buoy
9		Buoy with topmark (ball) (see L-70)
10		Barrel or Ton buoy
(La)		Color unknown
(Lb)	FLOAT	Float
12	FLOAT	Lightfloat
13		Outer or Landfall buoy
14	BW	Fairway buoy (BWVS)
14a	BW	Mid-channel buoy (BWVS)
†15	R "2" R "2"	Starboard-hand buoy (entering from seaward)
16	"1"	Port-hand buoy (entering from seaward)
17	RB RB	Bifurcation buoy (RBHB)
18	RB RB	Junction buoy (RBHB)
19	RB RB	Isolated danger buoy (RBHB)
20	RB G	Wreck buoy (RBHB or G)
20a	RB G	Obstruction buoy (RBHB or G)
21	Tel	Telegraph-cable buoy
22		Mooring buoy (colors of mooring buoys never carried)
22a		Mooring
22b	Tel	Mooring buoy with telegraphic communications
22c	T	Mooring buoy with telephonic communications
23		Warping buoy
24	Y	Quarantine buoy
†24a		Practice area buoy
25	Explos Anch	Explosive anchorage buoy
25a	AERO	Aeronautical anchorage buoy
26	Deviation	Compass adjustment buoy
27	BW	Fish trap (area) buoy (BWHB)
27a		Spoil ground buoy
†28	W	Anchorage buoy (marks limits)
†29	Priv maintd	Private aid to navigation (buoy) (maintained by private interests, use with caution)

Buoys and beacons.

LIGHT SYMBOLS AND THEIR MEANINGS

Symbol	Meaning	Description
F	Fixed	A continuous, steady light
F Fl	Fixed and flashing	A fixed light varied at regular intervals by a flash of greater brilliance.
F Gp Fl	Fixed and group flashing	A fixed light varied at regular intervals by groups of 2 or more flashes of greater brilliance
Fl	Flashing	A single flash showing at regular intervals, the duration of light always less than the duration of darkness
Gp Fl	Group flashing	Groups of 2 or more flashes showing at regular intervals
Qk Fl	Quick flashing	Shows not less than 60 flashes/minute
I Qk Fl	Interrupted quick flashing	Shows quick flashes for about 5 seconds followed by a dark period of about 5 seconds
E Int	Equal interval	Duration of light equal to that of darkness
Occ	Occulting	A light totally eclipsed at regular intervals, the duration of light always equal to or greater than the duration of darkness
Gp Occ	Group occulting	A light with a group of 2 or more eclipses at regular intervals

rules that governs inland waters are the Inland Rules. And on some North American waters the local government dictates the rules, such as on the Great Lakes and certain stretches of the Mississippi (and its tributaries), Atchafalaya, and Red Rivers. You need to know, understand, and comply with whatever set of rules apply to where you are.

Now wait. If all this sounds confusing and is starting to put you off, let me assure you that maritime navigation and seamanship is not any more complicated or intimidating than driving a car in a city or on an interstate highway. It is a learned skill that can substantially extend your evasion possibilities.

I highly recommend taking the Coast Guard Auxilliary's Boating Skills and Seamanship course. Use another name when registering. Just call information for your local branch and they'll set you up.

Coast Guard Encounters

If you are approached and told to heave to by the Coast Guard, do so. They are generally armed and may fire if they feel threatened. If they appear unarmed, they still have radios and will call their buddies who have a bigger boat (or helo support) and *will* be armed. Bluff and cooperate. Have all your safety gear and false identification ready. Be friendly, lie well, and good luck.

BOAT MAINTENANCE

We know that the less contact you have with people, the better. Therefore, the more competent you are at maintaining your boat, the less you will have to depend on others to maintain or fix it.

This past summer I was traveling up north by boat. The outboard motor I was using was about 15 years old, and one morning it just wouldn't start. This was bad, since I had to get going right away. I pulled off the shroud and found the prob-

The engines of a walkaround.

lem was the choke, which had become worn with age and was pulled farther out than where it was supposed to go. I reset the lever and knob and got under way; a simple problem requiring a simple solution.

Outboard Motors

Outboards have come a long way since I operated my first one in the Everglades back in the 1960s, a Chrysler 9.9 horsepower motor that was very quiet but a pain to have fixed because no one serviced them back then, or so it seemed. Since that time I have run what I sometimes think is every brand and size of outboard, all the way up to twin 200-horse Suzuki monsters attached to a Stamas 270 Tarpon. Over the years since my first outboard, motors have become more fuel efficient, easier to service, faster, and a lot more expensive.

As is the case with anything mechanical, preventive maintenance is absolutely the way to go. You start your preventive maintenance program by first reading the manufacturer's book on your motor from cover to cover. You might be surprised how few mariners do this. These books are filled with crucial information destined to greatly increase the life of your motor. Read it and follow its suggestions.

If you buy a new motor you will be told, hopefully, by the salesman that the motor has a break-in period during which you should never exceed a certain speed. This period will be given in engine hours, i.e., don't run the engine at speeds that move your boat more than 25 MPH until you have 20 hours on the motor. (This is an example; your motor might vary substantially.) An outboard isn't like an engine on a Dodge Viper, which will propel you right off the lot at warp speed with four miles on the car's odometer and not bat a cam shaft. You must baby them at first and continually check for parts that are working their way loose and fluid levels that are running. Recently I was in a new boat with a new motor and a warning beeper kept going off. The operator kept ignoring it at first, thinking it was just a glitch in the system, but I advised him to stop and check. He did and found that the engine's cooling system was failing; water was not running from the lower unit into and around the upper unit to cool the motor, which would have soon destroyed the entire motor. It was a quick fix.

You must check the fluid levels daily. The oil can drain out in minutes if a leak is acquired, and once the oil is gone, the motor is gone. Older motors and even a few new ones that are on the smaller side may require that oil be added to the gasoline directly, and you have to know the ratio. Most modern outboards, on the other hand, have internal oil reservoirs. Check daily just in case the warning beeper isn't working right. Never trust a beeper to tell you everything.

Check the cottar pin that holds the prop on the lower unit. Have extra pins aboard at all times. A spare prop is a must, too.

Fouled plugs are commonplace. Keep extras aboard and know how to change them quickly. You will need the right tools to do this.

Fuel lines also clog. When refueling at a marina, keep in mind that you don't know where the owner bought the gas or how long it has been in the storage tank. Gas can gum up faster than you might think. Always add a gas treatment to your gas to help prevent this problem.

Inboard/Outboard (I/O) Motors

There exists on Kennebago Lake a beautiful Chris-Craft many years old. She is made of wood and gleams in the sun whether she is running on the lake or resting beside her dock (where there sometimes is a large bull moose keeping her company). Her inboard/outboard motor is as meticulously maintained as her hull. Her very existence is clear evidence that proper maintenance will extend the life of an I/O indefinitely.

I/Os require just as much attention as outboards, more in some cases. They are more difficult to work on because there is more to them and they sit *in* the boat as opposed to *on* the boat. Still, some boats require them because of their weight and design.

WATER ROUTES

The water route you select can be critical. You want one that is either highly traveled so you can blend in or one that no one would ever consider. There are many that satisfy both requirements. What follows is a partial list of possibilities.

An Aquatic Highway

Running from Pennsylvania to Key West, the Atlantic Intracoastal Waterway is one of the best assets available for the fugitive on the Eastern Seaboard. It allows him to travel thousand of miles with almost no chance of being stopped by the authorities, except perhaps for the occasional Coast Guard safety inspection. Get on it and go.

The waterway is literally a highway, one used by untold numbers of people throughout the year. Well maintained and very safe, it offers everything you will need to make a go of it: food, fuel, water, medicine, clothing—everything. Anchorages are everywhere and many are free; no need to pull into a marina and be charged an exhorbitant fee to moor your boat for the evening. Just drop anchor in one of the myriad bays, sloughs, and coves that are maintained for that very purpose.

The Pacific Northwest

The Pacific Northwest offers tremendous evasion possibilities for the waterlogged fugitive, especially in maritime British Columbia. This remarkable labryinth of intricate channels, bays, coves, and rivers is perfect in many ways. There are many coastal ports in this region, but you must have a shipshape craft and abide by all the rules, since water cops are everywhere.

One thing you must be especially alert for in the Pacific Northwest is foul weather. The water is always cold up here, so if you go overboard or otherwise end up in the water, hypothermia is going to be a problem. You will need an exposure suit, which aren't cheap but which are a necessity in these waters. If you do not have or can't afford such a suit, then a watertight river bag (available in any larger sporting goods outlet) containing an extra set of warm clothes, high-energy food, and firemaking supplies is a must. If you can make it to shore with this bag before passing out from the cold water, you will have an excellent chance of survival.

The Great Lakes

The Great Lakes—Superior, Michigan, Ontario, Huron (including Georgian Bay), and Erie—are more like small oceans than lakes. (No, regardless of a moronic Congress listing Lake Champlain as one of the Great Lakes because it was politically correct at the time to do so, history and geography tell us that this lesser lake is not one of the Great Lakes.) The fugitive can evade on these lakes forever if he's careful, but they aren't perfect, since they offer all the same drawbacks as other large bodies of water, those being storms (Gordon Lightfoot wasn't kidding), cold water three months out of the year (including ice flows), and a lot of Coast Guard (U.S. and Canadian) and other bothersome people out on the water checking on you. Of course being lakes, they also leave you "water locked" in the sense that, much like being on an island, your escape routes are limited. Any inlet or outlet is going to be a security checkpoint.

The St. Lawrence Seaway joins Lake Ontario to the Gulf of St. Lawrence between Quebec, Newfoundland, and Nova Scotia. This allows you access to the Atlantic Ocean and the Canadian Maritimes as well as the Atlantic Intracoastal Waterway further to the south.

When operating on any lake, you should have a depth-sounder on board. You would be amazed at how easy it is to run aground, even when you think you are in deep water. Never trust charts to tell you *exactly* how deep it is in a certain spot, because depths can change radically, especially after a storm. Inlets, rivers, and sounds can see quick depth changes even without a storm.

GOING FEET DRY

Now let's have a look at getting on the road. This section will be comparatively brief because I have covered most aspects of land travel in my previous two books.

On the Road Again

Living on the road as a fugitive certainly has its advantages, and it is an option you might want to consider. A small RV that is mechanically sound and blends in well with the other RVs on the road and in campgrounds is the way to go. Why small? Because RVs, even new ones, gobble gas like it's going out of style, and money is probably going to be a concern to you, unless you robbed the Luxor's gaming tables in Las Vegas, which is incredibly dumb in the first place since the boys who own those tables are funny about people stealing their money.

Buy a used RV from a private owner, not a dealer. When you buy through a dealer you have more traceable paperwork to fill out, including financing in many cases. Instead, pay cash for a small used one and use a false ID when transfering ownership. Be very cautious with false ID, since it may be easily traced, depending on its source. A case in point: today's newspaper had a story about one Michael Scott Congdon, a college student who turned himself in to the proverbial authorities who were conducting an investigation of a rash of fake driver's licenses from two states that had been showing up all over town. The cops searched his apartment and found his computer, which the big dummy had used to record bogus license information on more than 200 people he had sold licenses to for $125 each. Yes, the fool kept the info on the hard drive. Now the cops know the name used on each false ID that was created. The moral of the story is that you should create your own false ID if you need one; never buy from someone else.

RVs give you the opportunity to stay on the move on our highways and side roads, but the disadvantage is that the more you move, the more people see your RV, whether they are paying attention or not. Therefore to reduce this risk, stick to secondary roads instead of interstates and other highways. Many atlases are OK for finding secondary roads, but the best are those made by DeLorme Mapping of Freeport, Maine. DeLorme makes a series of atlases that go by state called the *Atlas & Gazetteer* series

Abandoned farms are commonplace, especially in the Midwest. They can make excellent one-night stays. This one was found just outside a small Minnesota farming community.

(they have them for more than half the country and are always publishing new ones). These neat references show in great detail all the back roads, plus everything you would want to know about an area, like trails, passes, parks, Bureau of Land Management (BLM) land, state and national forests, and other recreation areas, plus all sorts of terrain features.

RVs have another disadvantage: they are easily broken into and are frequent targets of bad guys looking to make a quick buck. (The locks are easy to pick. I have a lock-picking friend who can get into an RV faster than a sitting president on a chubby intern.) The answer is to get a dog that looks, sounds, and *is* mean and keep him in the vehicle. Train him to growl and bark when someone rattles the doorknob or looks in a window. Very few burglars will still attempt to get in when a big dog is telling them that it is a good idea to stay the hell out.

Nevertheless, hide things of value just in case. Most burglars will stay in an RV less than two minutes, which means that the more difficult things are to find, the more nervous the crook grows with each second spent searching.

Train Travel

Contrary to popular belief, more and more goods are being shipped by train every year in North America, although passenger service remains stagnant. Freight trains are often a good means of getting from here to there, provided you know how to get on a train that doesn't want you on it in the first place.

Hopping a moving freight is exceedingly dangerous but often very simple. It is dangerous because if you slip, you can be crushed. This is bad. It is simple because, with basic precautions, you can get on, stay on, and get off with no fuss and no muss.

Here are some simple but important rules for hopping a freight train:

- Always get on at night. Daylight means that you are likely to be seen by someone, and that someone might just be socially responsible enough to report what they saw.
- Be prepared to defend yourself the second your hand touches the train. Rail riders already in the car you are trying to get on may see you as an easy mark for robbery.
- Never try to catch a train that is moving so fast that you

are having a hard time keeping up. One slip and it's curtains.

- Plan your route. Check the ground beside the tracks for obstructions. Again, one mistake . . .

Getting off a train is another thing. Never wait for the train to stop. Get off as it slows before it comes into the yard. Leave the area immediately; don't stay near the tracks. Be ready to roll when you hit the ground. Try to use a fall like a parachute landing fall, which means that you hit on the balls of your feet first, twist your body to the side with your arms tucked into your side, chin tucked, and legs together. Think like a banana; make your body roll onto the ground like a banana would, that is, curved. Let your momentum carry you rather than try to stop mid-roll.

I doubt you'll be carrying a motorcycle helmet in your E&E bag, but even pulling on a thick wool hat when attempting this maneuver may help cushion any blows to your head should it thunk on the ground in mid-tumble. You can also pad such impact points as shoulders, elbows, and knees with extra clothes stuffed inside your pants and shirt.

AIR TRAVEL

Generally speaking I don't advise unauthorized air travel, and today's security measures at airports are tough, for the most part. If you must, use a good false ID and make sure you don't have anything that will upset security when you go through. I recently forgot several items in a carry-on bag while going through an airport and nearly had a major problem. Fortunately, some things were confiscated with no charges being brought and others weren't recognized as being illegal by the security people (two switchblades).

I do not recommend hijacking a plane—small, medium, or

large, private or commercial. Hijacked planes are easy to track due to flight plan deviations and excellent radar in North America. There are much better ways to get around North America, so skip the movie stuff.

CHAPTER THREE

International Travel

> "Devil on the deep blue sea behind me,
> Vanish in the air, you'll never find me . . ."
>
> The Police, *Wrapped Around Your Finger*

One day I had to leave Turkey in a hurry. After traveling by car most of the night from a small village on the southwestern shore of the Sea of Marmara to Istanbul, I laid up for a couple of hours and then made my way to the airport. If you haven't experienced it, Istanbul's airport is precisely how you picture it: a malodorous madhouse caught in the riptide between Europe and the Middle East. I was booked on a flight to Berlin on Turkish Air and from there on Lufthansa to London. Out of London I had a seat on British Airway's Concorde to New York.

The first thing I did when I walked into the terminal was head straight for a ticket counter. I changed my reservations entirely—airlines, stopovers, and final destination. In fact, I went completely in the opposite direction, hitting Bombay, Tokyo, San Francisco, and finally Washington's National

Airport. I then went by car into the Virginia tidewater area and took a fast boat home down the Intracoastal Waterway.

I had no problems the entire way but may have had if I had stuck with my original travel plans. With due consideration for all the things that can go wrong, and with common sense and forethought, you can travel around the world with little trouble. One mistake or misstep, though, and you have a real problem.

Sometimes your past catches up with you when you would really rather it didn't. Returning to the United States from Mexico last winter, one of the people I was traveling with, a clandestine pilot and convicted marijuana importer, was once again stopped in Customs and given a thorough search while everyone else breezed right on through. It didn't help that he *looked* the part, and few of us were surprised when he was ushered into the back room to be gone over with a fine-toothed comb. He is no longer in that line of work, so he was clean and they grudgingly let him go.

The point is, you never want to look like your are someone interesting. It is bad for business. The late William Colby, former Director of Central Intelligence (DCI) who died on the Chesapeake while paddling his canoe, once remarked that he was the perfect spy because he could walk into a crowded room, mingle, and then leave without anyone really having taken notice of him. Strive to acquire this ability.

The thing about U.S. Customs is that you just never know. For instance, I recently returned to the States from Central America via Houston. I was doing some smuggling of things the U.S. government unreasonably says aren't allowed in the country—Cuban cigars—and was fairly certain I wouldn't be searched since I didn't declare anything on my declaration and said that I had been in Central America as a tourist, which is precisely what most Americans put on their declarations. I looked just like all the other tourists, too. I came into Houston during a busy time of day. As expected, the officer barely glanced at my declaration and waved me right through.

Nevertheless, I could have been the subject of a random search, which all customs agencies conduct, and you must be prepared for this eventuality. (For terrific insider information on the reality of borders and border crossing in and out of the U.S., I recommend *Beat the Border*, available from Paladin Press.)

OVERSEAS WATERS

Few fugitives ever stop to think about one of the most viable and clever means of evasion: using overseas waters to slip away and vanish. I have spent years at sea in foreign waters and can assure you that the possibilities are endless.

Some parts of the world offer excellent waterways for evasion purposes, whereas others have absolutely nothing to offer. Let's begin with Europe.

Europe

I have spent a great deal of time in Europe and have found water travel there to be quite easy and advantageous in maintaining one's privacy and freedom. However, you do have to know what you are doing. There are two types of water travel in and around Europe: canals and rivers and the ocean and seas.

Canals and Rivers

Europe's great rivers—with names like the Seine, Rhine, and Danube—are highly recommended for water travel, as are the excellent canal systems like the Canal du Midi, which links France's Meditteranean coast to its Atlantic coast by connecting with the Garonne River, offering a safe and fast passage between southern Europe and western Europe.

Using Europes canals and rivers allows the fugitive to move from country to country without ever having to face an angry sea. From southernmost Europe at the Italian boot to northernmost Europe at Germany's Nord-Ostsee Kanal, which cuts across the peninsula from the mouth of the Elbe at Brunsbuttel to Friedrichsort, you can travel at will—provided that "your

Navigation on European rivers means watching for obstructions like the gravel bar at right.

papers are in order," as well as your boat. You see, although these days it can appear that Europe doesn't even have borders anymore, the truth is that every country has its own funny way of wanting your papers and boat to be shipshape. Germany is especially picky on these points, but if you have everything in order—all the proper permits, boat documentation (safety and environmental), and a passport—you won't have a problem.

It can be easy to play the part of a tourist while moving through Europe. Visit this castle in Germany and everyone will think you are just another ugly American.

How do you learn about all the different regulations? Check with an experienced travel agency. If they seem confused or intimidated with getting you that info, go to another agency. Make sure you get up-to-date regulations before setting sail. And be sure you get the details of navigation on European inland waterways. Failure to do this might mean your running into a huge barge on the Main (pronounced *mine*) at the locks near Sachsenhausen, or it might find you suddenly heading out into the North Sea in a winter storm because you zigged (took the right fork) when you should have zagged (took the left fork) back at Zevenaar. But don't be intimidated! Many Europeans and Americans enjoy traveling on European inland waterways and have no trouble at all, so do seriously consider this option.

What's that? Just where are you supposed to get a boat to do all this? European houseboats, which are low-slung, somewhat narrow barges that have all the amenities of home and are quite quaint and welcoming, are affordable and available. They are available through advertisements in the local papers and in yachting magazines based in Europe.

Ocean and Seas

The Atlantic Ocean rubs up against Spain, Portugal, France, and the United Kingdom (England, Wales, Ireland, and Scotland). This is a lot of coastline for you to hide out along.

These coasts are filled with wonderful legends, lore, and stories of people who have made their lives by the sea for many centuries. As you might expect, pirates, brigands, and folks of a similar vein all made their livings in nefarious ways, and some are still prone to this lifestyle, which can work *for* you or *against* you, all depending on how you play your cards.

For instance, last year I found my way to Cornwall, a region well known for its colorful maritime history and in particular for the Cornishmen who would set signal fires along the shore at night in order to draw unsuspecting ships up onto the rocks, where the vessel would founder, the crew would perish (either in the "accident" or at the hands of the waiting Cornishmen), and the ship's holds would be emptied before you could say *aye, matey*. While I was there, a cargo vessel ran aground and was abandoned by her crew. True to form, the descendents of those brigands swooped down upon the ship and emptied her. Hundreds of doors were part of the cargo, and soon every home in the area had a new front door. It was quite amusing, and the

The Cornish coast.

police did nothing to stop it because a law on the books says once a ship is abandoned, she is fair game. It's called salvage.

The United Kingdom offers many little out-of-the-way locales where the fugitive mariner can come and go without anyone wondering about his business. Get in, get what you need, be cordial but speak to as few people as possible, and get out. No pub calls. Such places include three groups of islands off the coast of Scotland: the Shetlands, Orkneys, and Hebrides. Each of these groups have ancient maritime traditions and are therefore fully equipped to service you and your boat. One problem is that they are located in northern climes, especially the Shetlands, which consist of around 100 islands with only 19 of them being inhabited at all. You have to be an exceptional seaman to stand up to the tall water the North Atlantic can throw at you. Lerwick is the center of government and services.

The Orkneys are just north of the mainland and are separated from it by the Pentland Firth. There are about 90 islands in this group, and 30 of those are inhabited. For the best services, put in to Kirkwall.

The Hebrides are actually more like two separate groups consisting of 500 islands, less than 100 of which are inhabited. The Outer Hebrides are more northerly and westerly, the Inner Hebrides lying closer to the mainland. Carloway, Lochmaddy, Lochboisdale, Tarbert, and Stornaway in the Outer Hebrides and Port Askaig, Port Ellen, Ardlussa, and Tobermory in the Inner Hebrides are the better ports.

Try to pay attention and listen carefully when in these regions because the Scots have the damndest accent. And beware of storms and rough seas, especially off the Butt of Lewis, which is the northernmost tip of the main island in the Outer Hebrides, the Isle of Lewis. (I hope no one ever names a piece of land after my butt.)

From Spain to Denmark there are hundreds of ports available. However, as you go farther north you will find more and more ship traffic and frequently bad weather. From the French coast to Denmark is particularly bad, with the Netherlands being the worst, in my opinion. Be wary and alert.

Europe's Meditteranean coast is equally interesting to the waterbound fugitive and offers much better weather conditions. The drawback is that there is a much more developed tourist industry there, starting at Gibraltar and running all the way to Turkey. (Although it may be easier to blend in while traveling in high tourist areas, this can be a drawback because of the risk of being spotted and recognized by pure coincidence by someone on vacation. This may sound nearly impossible, but it happens with weird regularity.) This stretch of real estate includes world-class tourist and jet set destinations like Mallaga on the Costa del Sol (sun coast), Palma de Majorca (the city of Palma on the island of Majorca in the Balearic Islands; have a beer at the British pub on the hill above town), Cannes and Saint Tropez on the Cote D'Azur (home of the exclusive French Riviera; the yachts tied to the quay at Antibes are spectacular), and the Italian Riviera starting just east of Monaco and running to about Savona. The farther down the Italian coast you go the uglier it gets (more heavily developed and industrial), and this might be a better evasion area because of the nature of the place. The Italian police, however, have very fast "cigarette" boats, first made famous by cigarette smugglers around Naples. Give some thought to the island of Malta; they go out of their way to attract foreign money and interest.

Once you round the tip of Italy's boot you come into a nasty region that used to be Yugoslavia. It is rife with thousands of islands but dangerous because of the lawlessness of the region. Keep going.

Greece offers plenty of islands and out-of-the-way anchorages and ports, as does Turkey, but be very careful around Turkey. The Turkish police are as dirty as you have heard and have all sorts of little tricks up their sleeves to get you in major trouble. One common ruse is for a cop dressed in street clothes to ask you if you have anything to "trade," like a Zippo lighter or watch. If you sell him anything or make a trade you go to jail (that's right; they consider it all the black market, even for everyday items like pens or lighters), and it takes every penny you have to get out. I have had these cops try this very stunt on me in Izmir and Istanbul, but I had been around the block and knew the deal.

Izmir is a very cosmopolitan, sprawling city with a million alleys and hideaways. Towering skyscrapers overlook wretched ghettos, and businessmen wearing $2,000 suits stroll among street vendors dressed in little more than rags. Europeans, North Americans, Arabs, and many others all come to Izmir for the adventure of it and the business that is done there.

Istanbul is the same only much bigger. This ancient city is great for leaving false trails because it is a major crossroads between Europe, the Middle East, and Asia—leave the right ones and you could send investigators fruitlessly chasing you to Budapest, Cairo, and Calcutta. However, it can be very expensive, so be prepared.

Africa

Pay close attention.

Africa is one of the most dangerous continents for the boat-bound fugitive. There is no country I would drop anchor in, and that includes more advanced nations like Kenya (big problems there at the moment, with riots and general unrest) and South Africa (crime is out of control in many areas). From Alexandria to Cape Town, stay away. There are genuine pirates in many of these waters, including off the Horn of Africa where the island of Socotra sits. This island will get you killed. The pirates infesting it will shoot you mid-sentence, toss you to the many sharks circling your boat, and sail off with everything you once owned. Believe it.

The Middle East

Lebanon is now much more stable than it was in the 1980s, with the exception of southern Lebanon near the Israeli border. Stay well away from this region. Beirut, on the other hand, is again becoming a fascinating and versatile city now that the war is over. The port is open and it is business as usual, with thousands of vessels coming and going for all kinds of reasons. Everyone it seems speaks at least some English, plus French, German, and Spanish.

Israel is OK except for the fact that you must always be

alert for terrorism. Haifa is a major port north of Tel Aviv and has everything you need, but Israel is very expensive. Security is obviously a great concern of everyone in Israel, so everyone is armed and checkpoints are very commonplace. Israelis are an eclectic-looking bunch who come in all shapes, sizes, and colors, so you shouldn't have a problem blending in.

The Orient

The Orient is, in some ways, just as it has always been, with intrigue, adventure, and opportunity there for all who seek it. I have spent lots of time there—from Singapore and Thailand to Hong Kong, the Philippines, Japan, and Korea—and must say that I like it. But you have to be cautious.

For example, sailing around Japan is safe but for the typhoons and heavy shipping traffic. Head down to the Philippines and you have a somewhat different story. You are again at serious risk of typhoons—I once rode out a typhoon in the South China Sea back in 1978 and can guarantee you that it isn't fun—but now you also have to consider pirates again. Forget the image you have in your head about pirates that Hollywood put there when you were a kid. These guys are heavily armed and come in several boats that are fast and maneuverable. They are ruthless, just like their forerunners. (The book *Maritime Terror* has excellent information on modern-day pirate activity—and how to avoid it—in this region.)

The Philippines

The 7,100 islands in this archipelago represent everything from Nirvana to hell on earth for the fugitive mariner, depending on whether or not he runs afoul of the pirates and other hooligans thereabouts. Mindanao, the second largest island in the group, is very dangerous out in the "provinces," which is what everyone calls the countryside. Stay clear of villages. Watch for people who look at you with malice in their eyes and steer clear of them. Never talk to girls, no matter how pretty and friendly they are; you will be perceived by the men as a threat and you *will* have a problem. Be wary on trails. And here, as anywhere overseas, if you flash money like a drunken sailor, you don't belong in the fugitive business.

Davao, the largest city and major port on the island, is comparatively safer. I once got into a major shootout in these parts and, with some help from my compatriots, left 20 some odd dead bodies along the trail, so do be careful. The point is, if bad guys come for you in this region, they will come in numbers, well-armed, and ready to shoot on sight. You're not in Kansas anymore, Toto.

Tagalog is the most common language in the Philippines, but the official language is Filipino, which is based on Tagalog. There are more than 80 dialects spoken, but English is also commonly spoken and is taught in schools.

Instead of pulling into Manila, try Cebu City on the island of Cebu. This is a modern city (by Filipino standards), and the people are friendly. Another option is Olongapo, which is a port city north of Manila where the U.S. used to have a naval base. The mayor of Olongapo has done an excellent job of turning a disaster—the departure of the U.S. Navy—into a new beginning for his city, and all necessary services are again available for the seaborne fugitive.

Hong Kong

Hong Kong is up in the air right now. Having reverted to Chinese rule in 1997 after the British lost their lease, we are just going to have to wait and see what happens there. It used to be that you could sail into fabulous Hong Kong harbor and get whatever you wanted, both on the Hong Kong proper (Victoria) side or over in Kowloon. The periphery of Victoria Bay is still a huge floating marketplace, but Chinese rule may squash that. The Wanchai district of Victoria offers good bargains and adventure.

Caution is the key in today's Hong Kong. Never forget that

the Chinese think differently in many ways and that the communist government now has its eyes and ears operating all the time in all parts of Hong Kong. Be prepared to deal with officials; be respectful and never argue. Be ready for instant return visits from officials; just when you think you have gotten rid of a guy, the sonuvabitch will be back with six pals.

Vietnam

Vietnam is once again open for business. Ho Chi Minh City (Saigon) is as bustling as ever, and Hanoi and its harbor city, Haiphong, offer everything you need if you are in the north. As is the case with most of the Orient, there are lots of small craft about here, so be watchful.

All the cautionary rules for Hong Kong apply to Vietnam—times ten. The Vietnamese are extremely clever, treacherous, and absolutely unscrupulous. They will screw you in a heart beat. Your boat must be kept secure, too, because theft is extremely commonplace. And although tourism is booming, you never know who is watching you for what reason. Never forget that Vietnam is a communist country with entire agencies devoted to the tracking of foreigners within its boundaries.

Cambodia

Stay clear of Cambodia and its waters. Turmoil caused by Pol Pot, the Khmer Rouge, and the rest of the gang makes anarchy the most common commodity. Pirates are absolutely everywhere. If you see some boats coming your way with men standing on deck holding guns, open fire first and keep shooting until they break off or you die. Really.

Thailand

Thailand, on the other hand, has a lot to offer. You can anchor right off Pattaya and have small boats restock you for cash, preferably dollars. Don't leave your boat and go ashore, because it might be sacked while you are having a great time

The government security man at left is carrying a briefcase that undoubtedly contains a weapon. One mistake and you might find yourself dealing with the business end of that weapon. This is bad anywhere but worse in Vietnam.

with the girls (many of whom are HIV positive, incidentally). Another warning: if someone is chasing you at night down Pattaya Beach, beware of the ropes used to secure the boats to the beach. I have a scar on my right shin attesting to the fact that those ropes can be difficult to see.

Personal security in Thailand is as serious as it gets, with boats disappearing all the time and their owners coming up missing. Yanks are a common sight because of tourism and the heavy U.S. military presence in the area, so it's not all that difficult to blend in, but you have to believe that every Thai will take note of you and what you have. When you sail into port, you and your boat will be noted and scoped out.

Singapore

Singapore is a place of absolute law and order. It is beautiful, expensive, safe, and exotic. Beware of the ship traffic there and in the Straits of Malacca, which is one of the busiest stretches of water anywhere. Singapore is a city-state governed by the strictest of laws, and it is most unwise to break them. Even chewing gum in public is a crime.

The Sunda Islands

The Sunda Islands (Greater and Lesser) are the many large and small islands in the area between Borneo, Malaysia, Sumatra, Java, Timor, New Guinea, and Sulawesi. I've got to tell you, if I had to pick one remote region where I had to stay with my boat, this would be a likely candidate. This is one of those special regions where anyone can go and disappear off the face of the earth.

These are the tropics, and the many uninhabited or sparsely inhabited islands here make finding food and water and a quiet anchorage quite easy. The small ports often don't look like much, but ask around and you will almost always find what you need. Government officials and other suspicious sorts often are not about, and for the most part it is a safe area.

Do, however, use caution when landing on remote areas of islands because on some, like Flores and Komodo, there are large lizards called Komodo dragons that grow to 10 feet in length and are extremely dangerous. They feed mostly on goats, but they have been known to kill and eat humans. They wait along trails in the bushes and lunge out at whatever comes by.

Once an adult has you in its jaws, your chances of living to tell about it are slim.

Asia

Asia for our purposes includes places like India, Arabia in general, Pakistan, Sri Lanka, and so on. (The Orient is easternmost Asia.)

The most dangerous place for the waterborne fugitive in this neighborhood are the waters near Iran. Iranian Revolutionary Guards patrol these waters in gunboats and make general pests of themselves, making extreme caution paramount. I have spent lots of time around here and can personally attest to the dangers presented by these assholes, despite Iran just having elected a more moderate leader.

A good way to traverse these waters safely is to shadow a U.S. warship. Contact them first and ask if it is OK for you to follow fairly closely (but not too close) so that you don't have to worry about Iranians with fast boats and guns. All you have to do is ID your vessel and state what you want. They almost never ask detailed questions about you; they just want to make sure you don't endanger their ship, and they will almost always grant permission.

This region is always having some sort of a problem, so be ready to bag it at a moment's notice if trouble comes calling. As you can see by the photo on page 81, war is always iminent here; I took this photo in March of 1991 over Kuwait Bay, Kuwait City's waterfront. Obviously, boats of all kinds were targets of Saddam's henchmen. The owners all knew that Saddam was massing troops along the Kuwait-Iraq border but still failed to evacuate before it was much too late. Bloated bodies decorated the surface of the bay, adding a maccabre flavor to the surreal scene. The odor was none too pleasant either, I promise you, and even the plentiful sharks got full (or disgusted) after a while.

The nations in this region are sticklers for having permission to enter their waters or harbors. It's not that it is difficult

Kuwaiti boats sunk by marauding Iraqis in 1990.

to get permission, but you do have to have it to avoid pretty serious complications.

India offers lots of coastline and with it many ports of call of all shapes and sizes. You absolutely, positively need to have all your immunization shots before stepping foot into India. This is the Third World by far and away, even though it used to be a British possession. English is spoken by many, but crime is ever-present, as is disease and pestilence. As is the case everywhere in the Third World, purify all water, including bottled water that looks like it might in fact be locally produced and bottled. Hepatitis and cholera, among other nasties, are everyhere. The fugitive needs to have and use a microfilter that takes out the bad guys down to at least five microns, preferably even smaller.

India owns some islands of the coast of Burma and Thailand that offer the fugitive easy access: the Nicobar Islands and the Andaman Islands. Port Blair in the Andamans is your best bet.

Sri Lanka, formerly Ceylon, is still having problems with the Tamil Tigers, who are terrorists. Stay clear.

Australia and New Zealand

Australia and New Zealand are modern countries with all the amenities you need. New Zealand, on the other hand, and to a lesser degree Australia, is adamant about knowing who is visiting them, so have your papers in order and don't fool around. Use caution in the waters off northern Australia, where some of the world's most dangerous and aggressive crocodiles live. The advantage of Australia and New Zealand is that tourism is big thereabouts, and expatriate Yanks are quite common. Still, boating rules are strict, so be sure to obey the rules and have your passport and other documents ready.

Oceania

Paradise.

Oceania consists of the islands of Micronesia, Polynesia, Melanesia, and technically Australia and New Zealand, but the latter two are very different from a fugitive's point of view (they are included because of the genetic makeup of their original inhabitants: the Aborigines of Australia and the Maori of New Zealand).

Polynesia includes Hawaii. Hawaii represents, strangely enough, an excellent place to drop anchor and resupply. Yes, I know it is a U.S. state, but you can sail right into any harbor in Hawaii (except for Niihau, which is an island set aside for native Hawaiians and is private property under the most strict access), including Hilo (on the "Big Island"), Kaunakakai (a small coastal town on Molokai with the friendliest people you will ever meet), and Kahului (Maui) and go ashore and buy anything you want with no questions asked. No one will ask for any form of ID.

Besides Hawaii, there are plenty of other islands within Oceania, including celebrated isles like Fiji (the main port is Suva), Tahiti (the main port is Papeete; perhaps the most spectacular island in the world), and Espiritu Santo. (I know a fellow who deserted from the Foreign Legion while on Tahiti. He liked the island but had to get gone fast because he was being framed for something or other and they were going to toss him in the klink or maybe even kill him.) There are far too many to list here that the fugitive could make use of, but some groups to consider are the Phoenix Islands (Republic of Kiribati), the Palau Islands (southwest of Guam), the Solomon Islands (east of New Guinea, northeast of Australia), and the Truk Islands (part of the Federated States of Micronesia).

A warning: Oceania has a way of drawing and holding forever the people who come there from afar, supposedly only to visit. You may never leave. For the best read on this neck of the ocean, pick up the late James Michener's Pulitzer Prize winning *South Sea Tales*.

Central and South America

Rich in history, Central and South America offer nearly endless opportunities to the fugitive. Still, caution is always called for, but more so in some countries than others. For example, Guatemala by far offers the most danger because of the government there, whereas countries noted in the 1970s and 1980s for their civil wars and internal strife like Nicaragua and El Salvador are safer and more stable now. Your best opportunities are Belize and Costa Rica. To get very away from it all, Nicaragua's Costa de Mosquitos (on the Caribbean side) is largely undeveloped and even forboding. You could really get lost in there.

You must be quite careful sailing around Central America, especially in large bays on the Pacific coast. The reason is the number and variety of jetsom floating around, such as entire trees, logs, branches, and other obstructions that can damage or sink your vessel. Rest assured that you may sink before anyone comes to help you, since no Central American country has a good, fast, serious Coast Guard like the U.S. does.

South America can be a real winner. Venezuela, Brazil, Chile, Peru, Uraguay, Equador, French Guiana, Suriname, Guyana, Argentina, and even Colombia are all safely used by people looking to stay out of sight. Mind your Ps and Qs and you probably won't have any trouble, but if you do, it might be serious trouble, since these nations aren't exactly known for their pleasant police officers, except perhaps for Argentina. Incidentally, it is no myth that many fugitive Nazis headed to this region after the war, and many are still there. Sounds like a movie plot, but it is fact. This tells you something about the region's evasion potential.

The Amazon basin is navigable for hundreds of miles, and this includes many of its larger tributaries. It is quite safe, but you should always take precautions to keep it that way. Food and water are very easy to come by in there, especially fish, fruit, and vegetables, and once you become familiar with the region, gas and other supplies are obtainable with little problem.

When traveling in Central or South America, it would be to your advantage to have at least a moderate grasp of conversational Spanish. Typical tourist phrase books are a reasonable source of this knowledge, as are certain cassette tape home-study courses.

The Caribbean

The Caribbean is one of the American fugitive's closest refuges, and it comes quite highly recommended. A boat is the way to go down here, and with all the necessary precautions you can really make a go of it.

Islands are the Caribbean's greatest asset. The Greater (the Cayman Islands, Jamaica, etc.) and Lesser Antilles (St. Thomas down to Aruba) make up much of the famous West Indies. With so many islands to chose from and people who largely happily mind their own damn business, it is an excellent fugitive destination.

Take Antigua, for example. A typical island in the Lesser Antilles, Antigua (and its little brother, Barbuda) exists on money brought in by tourists. (They also make some of the best dark rum in the Caribbean.) It is no problem to sail into St. John's harbor, the main commercial port (English Harbor is the big tourist port), without any hassle, refuel and restock (especially at the open market near the waterfront), and be gone. If for some reason you want to hang out with those with a little more affluence, head into Lord Nelson's Dockyard. For an out-of-the-way anchorage, try Half Moon Bay.

An American option is the island of Vieques, east of Puerto Rico. Owned mostly by the government for military use, it does offer some quick options. A friend living on the island calls it "the land that time forgot." Vieques seems to have been passed by when it comes to development.

OVERSEAS LAND TRAVEL

It is likely that you will be traveling overland while a fugitive. Fortunately, there are many options and many different places to travel to and around. But for very few exceptions, you have the entire planet; in fact, those exceptions are locales like North Korea, Iran, and Serbia. As is the case with sea travel, land travel requires that you study beforehand, pay attention, and be prepared. Avoiding unnecessary risks is a big part of safe travel. You simply must think ahead and be able to think on your feet. (See Chapter 7 for general low-profile precautions to take while traveling at home or abroad.)

Europe

Yes, the borders are all but down, but careful thought is still required if you wish to maintain your freedom. I've spent years in Europe and know it like the back of my hand. Nevertheless, I carefully plan my travels in order to avoid problem people and places.

Everything changed when the "Iron Curtain" came crashing down in the early 1990s. The Berlin Wall is history now, and Europe continues to move forward toward the European Economic Community (EEC) and even a single currency, which will help some countries, such as Portugal, and wreck others, like England. Things will undoubtedly change greatly in the coming years.

On the Road

Europe has excellent roadways that allow quick and easy travel. You must be familiar with European driving habits and rules (as well as road signs). Note, however, that many European cities were laid out before the advent of the automobile, which translates into maddening gridlock at times. It is *not* a good place to be trapped in a car.

In Germany, for instance, they have *autobahns*. These are superhighways with no speed limits except for at certain junctions. Germans grow up driving very fast and expect you to be just as fast or get out of their way. The left lane on an autobahn is for the very fastest cars. If you are doing a paultry 160 KPH (100 MPH), you are going to either highly annoy the German behind you or cause a serious accident; maybe both. The left lane is more or less reserved for cars going at least 130 MPH, preferably more—much more. Mercedes, BMWs, Porsches, and Audis own the autobahn and cruise at shocking speeds in the left lane with their left turn signal (and often their headlights) on at all times. This means to get out of their way—*now.* If they flash their lights at you, then they are really pissed off and are annoyed at your rude driving habits. *Move over.*

If you practice, you will be able to blend in well in Europe.

Another little rule in Germany: left turns onto another road or street are generally forbidden unless specifically authorized by a road sign. The reason is that the high-speed cars and their drivers coming up behind you don't want to have to slow down while your slow ass makes a turn. Seriously.

Another thing: you are expected to be able to handle your liquor and drive. Unlike America where alcohol is strictly regulated when it comes to consumption age, alcohol in Europe is considered nothing special (that's why they don't have the problems we do). If you want to have 10 *pils* (higher-quality beers) at the local *gasthaus*, or perhaps even a dozen shots of *apfelkorn* (a tasty but powerful liquor served chilled in a shot glass) and then drive home, go right ahead; just don't get in an accident. If you do, you are done for because you have shown yourself to be unable to drink and drive—a social *faux pas* of the worst kind. In many European countries the penalties for drunk driving are more severe than in America. (Besides, if you are a fugitive, you should *never* be drinking and driving anywhere!)

All European countries have similar rules. In France you can drink two bottles of a decent Bordeaux and hop right in the old Peugeot and go home. Hit somebody, however, and you are in big trouble. In Italy you can quaf *vino* (there's a lot to be said for a nice Chianti) until the place closes, but roll your Lamborghini or fall off your Vespa and you are going to be hating life.

A driver's license from any country in Europe is honored in every country in Europe. American's can either get a license in the country they are in (rules for getting one vary widely) or they can get an international driving permit from AAA that is recognized throughout Europe. This is the way to go, since you can present AAA with an alternate licence to get an international driving permit and then renew it. The permit costs $10, and you will need two passport-sized photos. They are good for a year, but you can get a renewal form when you get the original and have a friend renew it for you (with two new photos) before it expires. Warning: have the photos done in a travel agency or other location that will give you the negatives. This way the negs aren't floating around, and you can just have another print run off when the time comes.

On the Rails

Train travel in Europe is as easy as it gets. Each country has local, national, and international rail service. Prices vary widely. For international travel, a EuroRail Pass is a good option. This ticket allows you to travel at cheaper rates for greater distances.

Learn how to operate the various ticket-dispensing machines throughout Europe that are right in the train stations. This way you avoid having to buy tickets from a person who might remember you. Good, up-to-date travel guidebooks will have instructions for each type of machine.

Many international trains have private compartments with sleepers available. This is the way to go if you can afford it. Otherwise, you will be sharing a compartment with other travelers. Check your tickets, though. I once traveled from central Germany to Roskilde, Denmark, and thought I had bought a ticket for a private sleeper compartment both ways. I hadn't. On the return trip I had to share a nonsleeper compartment, which was most annoying. And throughout Europe, be careful when buying a first-class ticket. It might be twice as expensive and the only difference is a slightly wider seat. This is especially true in the United Kingdom. On the other hand, fewer people will see you, since fewer can afford the ticket. It's your call.

Urban rail service may be in the subway style. In the U.K. it is called the "underground," in Germany the "U-bahn," in France the "Metro," and so on. Stations are clearly marked everywhere. Learn how to navigate the major European cities using these systems. They are usually very inexpensive and dependable.

National rail service runs from sometimes questionable (Italy) to absolutely dependable (Switzerland, Austria, and Germany), with everything in between. It is important that you figure out how to travel by these trains because they are a major and much-used source of transportation all over Europe. A good, simple plan is to watch everyone else and do what they do.

Before we move on to Africa, a word about youth hostels. Youth hostels are very common throughout Europe. They are inexpensive hotels with oftentimes communal sleeping quarters shared mostly by young people, but older folks use them as well. They are inexpensive, but privacy is out of the question and security is a concern, meaning you had better keep an eye on your gear at all times. Nevertheless, they remain a viable option.

Africa

Land travel in Africa can be anything from simple and safe to absolutely frightening. Accidents with cars are amazingly common and happen because of sheer ignorance and stupidity; likewise for trains. Never expect anyone to do the smart, safe thing.

On the Road

Africa is one of the most diverse continents on the planet, and driving conditions there are just as diverse. South Africa and Zimbabwe can be OK, but cross a border and you have to face a whole new story. Bandits, all sorts of wildlife in the road, crooked border guards, apparently deranged drivers (often drunk and armed), guerrillas, giant holes, and a general lack of road signs, long distances between gas stations, and many more challenges are present from one end of the continent to the other.

Driving around Africa calls for caution and a lot of common sense, and you must be ready for anything. For example, a friend of mine was crossing into Somalia one day (then called Somaliland) and was given a bit of a hard time by the guard. He ended up bribing the guy with a case of Chianti. Guess who the guard was? Mohammed Farah Aidid, the future warlord. With killers like these running around armed and in charge, you must be alert and ready to act quickly and decisively. You have been warned.

Animals like lions, elephants, leopards, Cape buffalo, rhino, crocodiles, and snakes (especially the mamba, which will chase you, is very big, has a very bad disposition, and secretes deadly venom that is often administered to the victim's head or neck, since the snake likes to raise itself off the ground to about that level to strike) take untold lives every year. The seemingly fat, slow, and stupid hippo takes more lives than any other wild animal in Africa.

One of the most daunting things about travel in Africa is the fact that many of the countries are factionalized. For example, the current (well, at least at the moment) Angolan military is called the Forcas Armadas Angolanas (FAA; Angolan Armed Forces). These guys replaced, in a manner of speaking, the Forcas Armadas para a Libertacao de Angola (FAPLA; Liberation Armed Forces of Angola). Joseph Savimbi's UNITA (Uniao Nacional para a Independencia Total de Angola is Portugese for the National Union for the Total Independence of Angola); the Frente de Liberacion de Enclave de Cabinda (FLEC; the Front for the Liberation of the Enclave of Cabinda—a hapless province torn by years of strife); the National Front for the Liberation of Angola (FNLA; Frente Nacional de Libertacao de Angola); and the Movimento Popular de Libertacao de Angola (MPLA; the Popular Movement for the Liberation of Angola, which is currently ruling Angola) are all potential problems for the fugitive who doesn't look and think like them, and you almost surely don't look and think like them. If you think their titles are confusing, wait until you try their various political agendas. And that's just Angola!

The most stable nation in Africa at the moment is Botswana (formerly Bechuanaland), where peace and democracy have been in order for decades and where there appears to be no trouble in sight. Zimbabwe (formerly Rhodesia) isn't bad either. Still, even nations with some development and a stake in the world community can be absolutely deadly. Take South Africa, for instance. You just don't go into downtown Johannesburg or Capetown. Crime in these cities and their suburbs is off the scale, with brutal murders being committed daily. Rawanda, Mozambique, Nigeria, Algeria, Ghana, Liberia, and Egypt all can be extremely dangerous too. Check the CIA and Department of State web site's for up-to-the-minute information and travel advisories. The CIA's web address is www.odci.gov; the state department's is www.state.gov.

On the Rails

African rail travel runs the gamut from luxurious (special safari trains that cost a king's ransom) to the more common—

dangerous trains driven by lunatics on *khat*, a narcotic plant that is chewed in the fashion of Vietnam's betel nut. You are running a risk whenever you board an African train, no matter what country you are in and where you are traveling to.

The Middle East

First and foremost, the Middle East must be thought of as a battlefield. It has been one since the beginning of recorded history. Treachery is a way of life; lies can be truths and vice versa. Extreme caution is what will help keep you safe here.

On the Road

Lebanon is mostly OK to drive around, but stay out of the southern part of the country near the Israeli border, where terrorists hang out and trouble is as common as graft and bribery in the U.S. Congress. Stay out of Syria. Jordan is OK, as is Israel, but stay away from the northern section near the Lebanese border, and stay out of Gaza and the West Bank for obvious reasons. Be very alert, and make your time in markets and on public transportation as short as possible due to bombings.

On the Rails

There is very little train travel in this region, and all of it is suspect. Count on the train being late. Count on it having few safety features. It will most likely be crowded. It will stop for no apparent reason. You will have your gear stolen if you take your eyes off it for a second.

The Orient

Land travel in the Orient can be very safe and very dangerous, depending on your planning, experience, and luck. Good intel about the region you will be in is very important. Before traveling overland in any Asian country, check with the state department or consulate/embassy and heed their advice.

On the Road

Driving around the Orient is generally safer than driving around Africa, but it still depends on exactly where you are. South Korea, Hong Kong, and Japan, for instance, are easy to get around and are safe. The Philippines, northern Thailand, and all of Burma (the Union of Myanmar, as it is officially known) can be either easy or threatening. For instance, driving around Manila is no different that driving around Tijuana, but in the Philippine countryside you have water buffaloes in the road, possible armed bandits, bad roads, mud slides, and other unique hazards. It all depends on where you are and when you are there. China is easy, but why would you be there? Cambodia? No way. Vietnam is easy around the cities but more demanding out in the sticks, generally because of a lack of gas stations and other amenities and water buffalo in the road. Also, you gotta watch out for bicyclists, as is the case throughout the Orient.

On the Rails

Oriental trains are wild and crazy. You can never tell if and when they are going to leave or arrive at their alleged destination. Generally speaking, urban trains are fairly safe; it is the trains that cross the countryside that tend to have major accidents from time to time because of bad tracks, bridges, and problems with the cars and locomotives. Speak with the consulate or embassy to see what they recommend.

Australia and New Zealand

Wonderful in many ways, Australia and New Zealand offer excellent land travel. Blending in, of course, is no problem insofar as facial appearance and skin color goes, but that accent can get you. Nonetheless, they are *still* discovering AWOL U.S. GIs from the Vietnam War living Down Under, which speaks volumes for its fugitive potential. Travel is very easy, with no checkpoints, roadblocks, and the like to worry about.

On the Road

Australia and New Zealand have superior road systems. Their citizens do, on the other hand, drive on the left-hand side of the road, which is something you'll need to get used to. Australia's rural regions, including the outback, have kangaroos that tend to cross or sit in the road at the most inopportune times. The red kangaroo, the largest, can demolish your car and kill you in an instant.

Another problematic little Australian critter is the tiger snake. This bad-tempered, fantastically deadly and aggressive snake commonly lies in the road at night. If you are walking along that road and step on it and it bites you, go ahead and kill it. It has killed you, so you might as well have some company on the way to hell.

On the Rails

Aussie rail systems are excellent and popular. Australia is a huge country (continent, actually) with three deserts: the Great Victoria, Great Sandy, and Tanami. If you are unaccustomed to deserts, then these are places you will want to avoid having to walk through. Train is a good way to move about, and no one asks for your ID once you have entered the country.

New Zealand is made up of two main islands, North and South. Again, strict entry control is the norm in New Zealand, but once in-country, train travel is available and dependable.

Central and South America

Land travel in this neck of the woods can be quite an adventure. It runs the gamut from pretty easy and dependable, such as in Argentina, to hazardous to your health. In Guatamala, for example, there are government death squads and hoards of bandits who like to ambush and rob you. Colombia has a wealth of drug runners and bandits in general. Belize has a lot of bad crime, and Peru has the remnants of the Tupac Amaru and Shining Path narco-terrorist groups.

On the Road

You will need to plan ahead to make sure there is going to be gas along your route. Extra jerricans of gas are a good idea, as is extra oil, belts, coolant, fuses, tires, a starter, a water pump if necessary, and so on. A gun is also a good idea.

On the Rails

Train travel varies widely in South America. In Argentina it can be marvelous, but in Colombia, Equador, Bolivia, and other less developed countries, you have to expect typical Third World problems with accidents, dependability, and so on. As always, the consulate or embassy can help with current conditions.

Asia

Insofar as travel goes, Asia is extremely diverse. For instance, in Bahrain it is easy to get around, but crossing the Saudi Arabian peninsula can kill you if you aren't prepared for sandstorms, extreme heat, and a complete absence of water, not to mention large camels standing in the road. In some countries you are going to find many checkpoints manned by bad people, but in others you will hardly see a soul.

You guessed it: ask the consulate or embassy before attempting to travel anywhere in this region.

On the Road

Driving conditions vary markedly across Asia. You absolutely must have good intel about the area or nation you are proposing to travel through. You are safe from crime for the most part when traveling around the Arabian peninsula in countries like Saudi Arabia, United Arab Emirates, Oman, Bahrain, and Kuwait, but Yemen and the People's Democratic Republic of Yemen are unstable. Iraq and Iran are, naturally, bad news, as is Afghanistan and much of Pakistan, but much of India is fine, except for Punjab State in northwestern India, where Sikhs may kill you for no good reason.

Russia is generally dangerous because of rampant crime and a deteriorating infrastructure. The nations that broke off from the old Soviet Union are mostly OK. China is mostly OK. Stay clear of Burma and Cambodia and the northern provinces of Thailand (a major poppy-growing region that is part of the Golden Triangle). Vietnam and Laos are generally OK except for government officials at all levels who are corrupt. South Korea is OK; you need not worry about North Korea, since you can't (and wouldn't want to) get in. Bangladesh is a mess. Tibet is beautiful but under Chinese rule, but Nepal is OK.

On the Rails

Extreme caution is called for throughout Asia when it comes to train travel. If the train and rails aren't simply unsafe because of maintenance problems, then crime will be the problem. India is the worst for maintenance and operation trouble, not to mention the occasional horrific bombing by domestic terrorists. You are risking your life.

And speaking of risking your life, never get on a ferry anywhere in Asia or the Orient. They sink with alarming regularity. Find another way.

The Caribbean

Although you can never drive very far in the Caribbean, you can sure get in plenty of trouble doing so. A friend of mine got into an accident on one island—he demolished a car with a bunch of Soviet government intelligence people in it, and they were at fault. Nevertheless, the island government wanted to put the arm on Bill just because. Bill didn't like that idea, so he left.

Caribbean driving rules are strange. For instance, on many islands if you run into a guy who was passing on an uphill curve at night, you are at fault because you should have known he was coming and taken measures to avoid him. No, really.

Chapter Four

Hideouts

> "Is there gas in the car?
> Yes, there's gas in the car.
> I think the people down
> the hall know who you are.
> Careful what you carry,
> for the man is wise,
> You are still an outlaw in
> their eyes . . ."
>
> Steely Dan,
> *Kid Charlemagne*

If you decide that remaining on the move forever isn't your thing, be it in a boat or a motor home, you will have to consider an interim hideout, which is someplace you can get to where you can stay for a while but probably not forever. Now, I am not speaking of a remote cabin in the woods but rather a country, possession of a country, island, or principality that affords you a life away from prying eyes and off the beaten path (somewhat or very) for a few weeks or months. An interim hideout should be fairly easy to get to and be home to people who don't really give a damn about who you are and why you are there. You should be able to blend in fairly easily. Customs officers and other government authorities should be incompetent, lazy, or easily bribed.

Some temporary hideouts can end up being safe permanent hideouts. It all depends on your situation and the

A vendor displays her wares in Belize City.

country you are considering. For instance, after hiding in the cloud forest of Costa Rica for a few months, you might come to think that heading into San Jose to take up permanent residence is OK, and it might be. You and your situation dictate this; there is no broad rule that applies.

What follows are just some of the places that might suit your needs. Some are obscure, others are not, but each place is a potential hideout for you. Remember, these are temporary hideouts; more permanent situations are covered in the next chapter.

BELIZE

Formerly British Honduras, Belize is part of Central America and lies just below Mexico's Yucatan Peninsula. Small and dependent on tourism for its livelihood, Belize's population is made up of quite a hodgepodge of people, from white to black to Mestizo to whatever. If you get your scuba diving instructor qualification (from the Professional Association of Diving Instructors, the National Association of Underwater Instructors, or similar organization) you may be able to secure work in Belize in that profession. There are several other tourism-related fields that could provide you with work, too.

Belize, however, can be quite expensive when it comes to things that we take for granted here in America. For example, a can of Deep Woods Off will run you U.S. $7.50, and a 54-quart cooler will set you back a couple of hundred bucks.

CANADA

I am constantly amazed at how many people fail to consider Canada as an evasion location, this despite the fact that English is spoken by most Canadians, most Canadians are of European stock, and it abuts the United States for thousands of miles, the border being easily crossed on foot or by boat or canoe. Canada may, in reality, be one of the best bets for a fugitive.

But Canada is a funny place with funny laws. For example, freedom of speech and freedom of the press aren't what they are in America. Did you know that you could be arrested for transporting certain Paladin Press books across the Canadian border? Did you know that Canada has some of the strictest firearms laws in the world? These are just some of the little things you need to be aware of when traveling up north.

Canada is very diverse geographically, with broad, wind-swept plains, towering mountains, coastal plains, Arctic tundra, woodlands, rain forests, and more. There are large cities with all the traditional social services of the developed world, and tiny enclaves and villages that seldom see a stranger. The weather ranges from warm to absolutely frigid. The forests are filled with animals and the waters are filled with fish.

In a nutshell, Canada is great as an evasion site because of the ethnic extraction of most of the people (white Europeans), huge land mass, varying climates, many jobs, not-too-bad economy, friendly people, and any number of other factors. Seriously consider it.

COSTA RICA

Costa Rica is the home of many genuine American expatriots, or "expats," who have made new lives for themselves in a big way. Central America's most advanced nation socially and economically, this country of triple-canopy jungle, cloud forests, mountains, mangrove swamps, and even arid regions is one of the fugitive's best relocation options. There are large and small cities, towns, and villages. I was last there in March of 1998, enjoyed myself immensely, and, as always, found several expat Americans purely by chance in such diverse regions as Zancudo on the south Pacific coast (he is running a lodge thereabouts) and downtown San Jose near the Central Market (he was running a book and cigar store). Every American expat I met was obviously happy and living a good life. You can too.

Depending on your circumstances, you may be able to drive to Costa Rica rather than fly or sail. A way to do this is to use the Inter-American Highway starting at McAllen, Texas. The trip, if all goes smoothly (and it probably won't), will take you 10 days from Texas to San Jose, the capital of Costa Rica. Travel only during daylight and use common sense as to where you spend the night. If you have experience at traveling through the Third World at night, you can sleep under the stars, but be aware of insects, snakes, bad water, and other unpleasantries, including bandits. Use a "Club" to lock your steering wheel. Be prepared to pay "special taxes" at every border crossing. (After

Downtown San Jose, Costa Rica.

Mexico you will hit Guatamala, El Salvador, Honduras, Nicaragua, and then Costa Rica.) Make sure your vehicle is in perfect running order, and have extra gas, water, coolant, belts, and a battery along, at a minimum. Never drink any water along the way that hasn't been purified.

HONG KONG

Although Hong Kong has reverted to Chinese rule, it

remains an excellent place to vanish, if not simply for the sheer numbers of people there. Yes, ethnic Chinese are clearly the dominant race, but with some basic changes to your appearance and an entreprenurial attitude, coupled with the Chinese tradition of seeing nothing and hearing nothing, you could really make a go of it in Hong Kong.

Hong Kong is divided into three general areas: Victoria, Kowloon, and the New Provinces. The latter is like the countryside, although the region is being built up rather quickly and has been under development since the first time I was there in 1978.

As is the case wherever you go, use great caution when speaking in public. A friend of mine recalled how he was once hanging out in a Hong Kong bar with a "round-eye" (Caucasian) who intentionally kept his mouth shut in that particular situation. The Chinese there had no idea the man spoke fluent Mandarin and Cantonese, and they were quite shocked when he became irritated at the goings-on and conversations in the bar and spoke rather harshly to the proprietor in perfect Chinese. The people in the bar quickly wised up when they realized that the Caucasian who just startled them was one of the highest ranking police officers in Hong Kong. Likewise, if you don't know who can understand what, then don't say anything you don't have to say, and never assume there is no one nearby who can understand you.

The best area to do your shopping in Hong Kong is either down on the waterfront in the massive and everchanging floating marketplace or in the Upper Wanchai district. Avoid the trendy and expensive skyscrapers downtown and over in Kowloon. If a merchant doesn't speak English, use pantomime to communicate. It's easy. For the best steak in town, hit the 747 Restaurant. If you want to go to a bar, go to one off of a side street and off of the beaten path. Skip the stupid tourist traps like Suzy Wong's.

I recommend these places because they see a diverse selection of visitors; people won't be stopping dead in their tracks to stare at you. And when it comes to blending in in Hong Kong, simply go with either the typical tourist look (dress as a normal North American) or the white Hong Kong businessman look, which is in a nice, dark suit.

If you need to get out of Hong Kong without going through customs, buy a seat on any merchant boat heading for Macao. Try to arrive at night, preferably when it is raining. Cover up and just walk off the boat when it docks on the waterfront. Act like you know what you're doing and keep your face covered by a hood or hat and no one will likely bother you.

THE SAN ANDRES AND PROVIDENCE ISLANDS

I'll bet that you have never heard of these islands, have you? I didn't think so. That's one of the reasons why they are a good choice as an interim hideout.

These islands are off the Atlantic coast of Nicaragua but belong to Colombia. They are used primarily for sport fishing and as narcotics transhipment sites. Spanish and English are the major languages, and the islands have all the amenities you need. Additionally, this chain is infrequently thought of when it comes to evasion and those hunting evaders. Although there are not that many good jobs waiting for you, if you have marketable skills that can keep you self-employed, you can do fine.

THE PHILIPPINES

The Philippines—or the P.I. to those who have spent a lot of time there—is an archipelago consisting of more than 7,000 islands. There used to be a major U.S. presence there, almost entirely on Luzon, when we had two major military bases: Subic Naval Station and Clark Air Force Base. Today, however, both bases are closed and the population of Americans is greatly reduced.

The vast majority of the islands constituting the Philippines are uninhabited or sparsely inhabited, and most are quite primitive, with no airport, little if any electricity, and all but no amenities. Mindanao, Mindoro, Cebu, Negros, Masbate, Palawan, Panay, Samar, Bohol, and Leyte are the other main islands. The terrorism problem presented by the New Peoples Army there is largely gone now that the U.S. has departed and Ferdinand Marcos and his shoe-hog wife, Imelda, are no longer in power, although may members of the NPA are now simply bandits roaming the countryside and preying on the helpless. However, the Moro National Liberation Front (the "Moros"), a Muslim separatist terrorist group operating largely on Mindanao, is active again and killing many innocent people. Best bet? Stay the hell off of Mindanao.

Three big advantages to hiding in the P.I. is the sheer number of small, uninhabited islands available, the fact that most people speak English, and the warm weather (typhoons can be a problem, however). The jungle (37 percent of the P.I. is forested) is full of food (hundreds of species of reptiles, a few species of deer, and 760 species of birds) and water, and it is quite easy to hide there.

Disadvantages? You have to know what you are doing in the jungle, because there are many ways to go wrong. Also, some areas of the Philippines are notorious for bandits and crooks, so you have to watch your ass all the time, especially around the southern islands where pirates roam the seas at will. You might find that the remote islands have people who speak little if any English, and there are hundreds of dialects spoken. Forty-three percent of the P.I.'s employed people are farmers of some kind.

VENEZUELA

Venezuela offers the fugitive a good interim hideout. There are possibilities for urban evasion in the capital of Caracas, and rural possibilities are equally good, with 40 percent of the country being forested and six navigable rivers, including the mighty Orinoco, for transit. There are many islands off the coast, too, with Margarita being the largest.

Venezuela consists of four clearly defined regions: the Northern Highlands, Guiana Highlands (with elevations reaching to 9,000 feet), *Llanus* (tropical grasslands), and Maracaibo Lowlands. Each of these regions has advantages and disadvantages. For instance, the lowlands have lots of snakes like the anaconda and the weather is humid, but there are many fish thereabouts. Still, the rural evader can find rich animal life everywhere, including sloths, a huge variety of birds and fish, bears, anteaters, deer, monkeys, and seemingly countless reptiles, including crocodiles (the Orinoco croc grows to about 21 feet).

A major disadvantage is the fact that you are a *gringo* in Venezuela, and the people might turn you in for a pittance if they get wind that perhaps you are being looked for. A good way to go down here is on a houseboat on the Orinoco.

CHAPTER FIVE

Foreign Citizenship and Living Overseas

> "They don't even have policeman one,
> Doesn't matter where you've been or what you've done.
> Do you have a dark spot on your past? Leave it to my man, he'll fix it fast.
> Pepe has a scar from ear to ear, he will make your mug shots disappear."
>
> Steely Dan,
> *Sign In Stranger*

If things are so grim that you have to leave the country and expect to never come back, you will probably have to change your citizenship. You will have to consider which country you want to live in and become a citizen of very carefully. Some things to consider are:

- What it is going to take to get citizenship there?
- What standard of living can you accept?
- What is the climate like?
- What languages can you speak or feel you can learn fairly easily?
- What marketable skills do you have?
- What are the people like (do you like them and can you get along with them easily)?
- How easily can you blend in, if that is important?
- How often and to what extent are you going to have to bribe local and national officials (if at all)?

- What are the chances that someone is going to come looking for you there (someone you would rather not find you)?

Before you start laying plans, you must seriously consider what you are about to do. Most countries require you to give up your American citizenship if you want to become a citizen of their country, although there are exceptions for dual citizenship, and in some cases you can actually have three or more passports. For example, I know one guy who carries American, Irish, and British passports. He is an American who lives part of the year in England and has a business there. His British wife is also an Irish citizen. This situation allowed him to get both British and Irish passports.

You will note that he didn't give up his American citizenship, although he could have if he wanted to. There are many substantial advantages to being an American, as we have rights that many countries do not have. We enjoy no small degree of safety and comfort, generally speaking. Our economy is doing well, and the nation is very stable politically. These things all add up, and you must be absolutely sure that changing citizenship is what you *must* do. Once you give up your American citizenship, there's no getting it back.

WHAT IS IT GOING TO TAKE?

Different countries have different requirements for citizenship. For instance, some demand that you live there for a while and perhaps have a job there, some want you to have a spouse there, some want you to bring capital into the country, some want you to bring a certain skill (doctor, chemical engineer, military skills, water purification technician, agriculture experience, etc.), and others just want a fee up front. You must do your homework if you expect to be successful.

The best way to look into this is by checking with the country's embassy, all of which are located in Washington, D.C.

Some countries also have consulates in major cities like New York, Los Angeles, Chicago, and San Francisco. Ring them up and ask to speak to someone who can tell you about becoming a citizen of their country. Ask the right questions (those listed under the "things to consider" list), and never give your true name. If it sounds OK, find out how the process gets started and take it from there.

STANDARD OF LIVING

In case you haven't been there, the so-called Third World is nothing like the United States or other developed countries like those in Europe. Before you head for East Llamabutt to set up shop, you must complete a thorough country study to see if it is definitely someplace you could live. I have spent many, many years in the Third World and have found comparatively few countries where I would feel comfortable and safe over the long haul. I wouldn't like central, northern, or western Africa, but I could do OK in East or Southern Africa, where there is less mayhem, disease, and pestilence on the whole. I could hang out in many places in the Orient, Indonesia, and Oceania but would have nothing to do with India and central Asia. South America is nearly all a piece of cake.

You must consider the details, such as whether or not you can deal with electricity that is about as dependable as the testimony of an FBI agent at the Ruby Ridge trial and local authorities about as trustworthy as a member of the U.S. Senate. How about running water? Is it important to you? You may or may not have it. Need a doctor? There may not be one for hundreds of miles, and if there is some sort of medical care infrastructure, it probably won't be what you are accustomed to. (Self-treatment is almost certainly going to take on new meaning for you.) Local grievances may not be handled by the government in the same manner they are here; you may have to take matters into your own hands. Communications services are also questionnable in most Third World countries, so you are going to have to deal with that, too. Need to put money in the bank? That bank might take your money and not be especially keen on giving it back when you want it. If you have dealings with the cops, they might not be anything like most cops in the U.S. Yes, you have a lot to think about before you move out.

Strangely, though, some Third World countries, although underdeveloped, don't have as many problems as you might think. Many island nations in Oceania and countries in the Orient (the Philippines and Thailand in particular) and South America don't have it so bad, although much of the population lives badly. Money makes all the difference in the world in such places. For example, if you have no money and move to the Philippines, you might end up in a hut in the jungle or a hovel on the outskirts of metro Manila. If you have some money—and it doesn't have to be that much—you can live like a king, comparatively speaking. The same goes for Thailand. Now, if you were a good photographer or writer (magazine articles, books, or both) and lived in one of these countries, you could quite easily earn a decent living and live well. On the other hand, if you end up in the squalor of the "ville" in Olongapo, you might find yourself dumping buckets of water over your head (your shower), squatting over another bucket (your toilet; which you then toss into the Shit River—that's the "round-eye" name for the Olongapo River, a giant open sewer system that runs through town), and having a tug-of-war with a huge rat (about the size of a beaver, I swear) over your underwear, which you so carelessly left lying on the floor (that never happened to me, really).

CLIMATE

Humans can adjust to any climate on the planet, but that doesn't mean they are going to like it. I have lived in temperate forests, mountains, deserts, jungles, rain forests, and many other places and have seen temperatures ranging from a bone-numbing -80° Fahrenheit to a broiling 130°. Pretty extreme, and I didn't like either end of that spectrum. However, I did survive and could do it all again if I had to, no problem. Nevertheless, I would rather not.

I have three friends who live in Alaska, one (Chris) way out in the boonies where you carry a large gun everywhere you go, one (Norm) in the coastal fishing town of Cordova, and one (Chuck) in Anchorage. Each feels comfortable where he is, but all could adapt to the other's lifestyle quite quickly. Sure enough, the key to climate is adaptation.

One very important thing to consider about climate is disease. Diseases more often thrive in humid, subtropical, or tropical environments than any other, so do give this some thought.

LANGUAGES

If you select a country where a language other than English is the most commonly spoken—and that goes for most of the planet—you must consider how long it is going to take you to master the the native tongue. Not all languages are created equal, either. Spanish, French, Swahili, Hawaiian, Tagalog, and German are ridiculously easy compared to Mandarin Chinese, Arabic, or the *click* language of the Kalihari Bushmen. And you really should learn the language, since that way you can listen to people speaking around you while they are talking about you and think you have no idea what they are talking about. Speaking the language also makes it much easier to deal with people culturally and practically. It tells them that you have gone out of your way to learn their language, which is a sign of respect that everyone likes.

MARKETABLE SKILLS

Chapter 6 discusses in detail some skills you can develop that will make you a marketable item over there, and you simply *must have* such skills if you want to eat, unless you ripped off the armored car you were driving and made off with all the loot. (If you did this, then you had better know what countries have extradition treaties with the U.S. For a current list of which countries have what types of extradition treaties with the United States, visit the state department's web site.) Yes, the chances are that you are going to need a job of some sort, so you had better start developing some marketable skills now, before you make a run for it.

Research is again important. Many developing nations are desperate for people with skills in engineering (all disciplines), medicine (all disciplines), and computer science, and many are willing to accept any piece of paper and line of bullshit you can give them that attest to the fact that you have the engineering skills of Leonardo DaVinci, medical expertise of Jonas Saulk, and computer software qualifications of Bill Gates. Bogus degrees, certificates, and resumés are astonishingly easy to get or make yourself, but you must have *some* skills in the field you are claiming to be an expert in, lest you be discovered for the fraud you are and thrown into some dank pit to rot.

Half of what's required to land a job is selling yourself. If you can convince someone that you are clearly the best and he just can't afford to let you slip by, the job is probably yours.

THE PEOPLE

There's an old saying that claims that people are the same everywhere. It was obviously made up by someone who went nowhere and did nothing.

People are absolutely *not* the same everywhere. It might not be politically correct or socially enlightened, but the truth is that some people are very bad and completely untrustworthy. They will take advantage of you, lie to you, steal from you, and sell you a bill of goods faster than you can say *Better Business Bureau*. Believe it.

Now, I am not referring to individuals here. I mean *kinds* of people. Let me explain.

It is an Arabic belief that if they got the drop on you or the better of you, then that's the way it goes; you should have been more shrewd or paid closer attention. This explains why Arabic history is so filled with factional fighting. It is Chinese and Japanese belief that the truth is in the eye of the beholder; if you believe it, then it must be true. This explains why the Chinese are so stubborn and merciless and why the Japanese still think the U.S. started World War II and victimized them at Hiroshima and Nagasaki. The Somalis are treachery and thievery in flesh and blood; they will kill you just because they can. Americans? We are great justifiers; if something bad has to be done, we will find a way to justify it, plain and simple, like the stealing of an entire continent from its rightful owners under the very questionnable principles of manifest destiny and eminent domain.

On the other hand, some people are far less likely to cause you a problem. The Kiwis (New Zealanders), Aussies, Irish, Scots, Finns, Germans (although they start world wars from time to time), Ethiopians, Belizians, and Argentines are among the most trustworthy and respectable, in my opinion.

Before you select a country, you had better study the people who live there. And never forget where you are. For example, it is improper to lump the Welsh in with British; call a Welshman a Brit and he may clobber you. Likewise, just as a Brit had better not call a man from Georgia a Yank, you shouldn't call a Spaniard Hispanic. Cultural sensitivity can not only help you maintain a low profile, it can keep you alive.

BLENDING IN

If you are in a situation where looking like everyone else is important, then do your homework and don't fall prey to baseless physical appearance stereotypes. Not all Italians have dark hair, some Spanish people (Castillian, not Hispanic) have blond hair and blue eyes, most Germans are not tall, blue-eyed blonds, many Zimbabweans and Argentines are white, and many Brits and French are black.

The best way to determine if you can blend in physically is to go to the library and read about the country in an encyclopedia, or get on the Internet and use a search engine to get information on its citizens. You can also use a CD like the Encarta Encyclopedia in your computer.

OFFICIALS

This is a touchy and tricky subject, as there are many misconceptions about bribery (what it is and what it isn't) and how it is done safely and effectively. In short, it can be very easy and almost out in the open, and it can be as slippery a slope as one can traverse. Caution is always in order.

First, bribery can be a serious catch-22. For instance, you might try to bribe your way out of a problem in Turkey and then get arrested for bribery. Once in jail, you must now bribe your way out, and it's going to be a lot more money than what you offered the cop in the first place. Three pals of mine had this happen to them in Izmir, so watch it.

Some officials are more prone to bribery than others, and it often depends on the country you are in, i.e., whether or not bribery is accepted there. Some countries and regions where bribery is rampant include Mexico (horrendously corrupt), all over Africa, Colombia, Panama, Russia, Vietnam, central Asia, and the Middle East.

The best way to bribe someone is to let them bring up the subject first. This isn't foolproof, since they might simply be setting you up, but it works more often than not. They will probably never refer to it as a bribe, so pay attention to follow suit. The official might take you aside in a room or some other place out of the sight of others and tell you what the "fine" is that you must pay. If you have it on you, pay it; don't deny having that much when all they have to do is search you. The "fine" might also come in passing pleasant conversation, such as, "Mr. Schmuck, that is a beautiful watch you are wearing." Translation: "Gimme that damn watch you piece of American shit!" Give it up. You can always get another watch. Again, *pay attention.*

If you feel that it might be worth the risk to broach the subject, do so carefully and casually, and use language that avoids the word *bribe.* Instead, make it seem like routine business and that you do not mean to insult the official. And make sure you offer enough! Be all smiles and pleasantries all the while. Make no demands to see a lawyer; this can get you shot. Remember, once you are outside U.S. borders, there ain't no such thing as the Bill of Rights. This is every bit as true in "Western" countries like France and England as it is in Singapore and the Sudan.

COMPANY'S COMING

Finally, you must consider the chances of someone from your old life coming to find you and the chances of being recognized by accident. If either of these is a real concern, then you must select a very out-of-the-way location. Fortunately there are plenty of them, but you must use the utmost caution to cover your tracks; any hint you leave behind that leads them down the right trail can cost you everything.

You must never return to a country you have been known to visit or have lived in before; that's the first place they will look. Leave a false trail, such as a computer memo to yourself (left on your computer at your last residence) or a note on a piece of scrap paper reminding yourself to pick up your ticket to Guatamala or to call Igor Svetlovics at the Russian Consulate. Of course you really didn't go to Guatamala, and Igor is the name of a Russian official at the consulate that you got by asking the receptionist who it is you should speak to about immigrating, when in reality you never planned to go there. These false trails can buy you loads of valuable time and actually work well enough to keep the bad guys off your trail forever.

You've got to have an evasion plan in-country, too. If you let your guard down for only a moment, that will be the very moment your enemies show up in your bedroom in the middle of the night. This could entail such items as a packed bug-out bag with clothes, medicine, and personal items; a current passport; cash, preferably U.S. dollars; disguise items; and even a loaded firearm if it's a truly dicey situation. Think like a Boy Scout and be prepared.

Chapter Six

Get a Job

"Agents of the law, luckless pedestrians, I know you're out there,
With rage in your eyes and your megaphones,
Saying all is forgiven, Mad Dog surrender.
How can I answer?
A man of my mind can do anything . . ."

Steely Dan,
Don't Take Me Alive

Unless you somehow came up with a great deal of money and have invested it wisely so that it is unlikely you will ever have to work again, you are going to need some form of income. What you choose to do for employment and where you choose to do it will depend on the sum of all your skills—physical, cerebral, and communication—and your situation, the latter meaning why you became a fugitive and where you feel safe living and working. For instance, if you ripped off the MGM Grand for a few hundred thousand dollars running an inside scam, you can forget about doing similar work in another casino anywhere in the world, and that includes Indian reservations and overseas in locales like Monte Carlo and the Bahamas. The people with vested interests in the casino you ripped off will come looking for you in every casino in existence. Them finding you will put your karma into a rapid and precipitous decline, not to mention your EKG.

INDEPENDENCE DAY

The best form of employment the fugitive can have is that which keeps him out of the public eye, away from full-time jobs—especially those that report him to the tax man—and provides him with enough steady income to live well enough. You might be surprised how many skills you can develop when you put your mind to it and use some common sense.

I am a perfect example of this. I have income coming in from dozens of sources because of my writing and photography and skills that have to do with weapons and survival. This doesn't mean that you must develop similar skills but just goes to show you that it *can* be done. You are strongly advised to *get* it done.

Let's take a look at some jobs that you can learn quite easily and get yourself started in with only a small investment.

Outdoor Photographer

I have a friend whose job it is to travel the world taking pictures of everything from railroad tracks in Australia to beautiful women sitting on the end of a dock in the Caribbean. His photographs have appeared in far too many magazines and books to name. I can't even begin to guess how much he makes annually, but it is a lot, and he is in great demand. He is remarkably talented and wasn't born with a single photographic skill; he learned everything. And that is the point: photography, like every other skill, is learned. You can learn it, too, and there is no need to rush out and buy $20,000 worth of cameras, lenses, filters, motor drives, tripods, and support gear. You can get started today for $1,000, probably substantially less. This is great for the fugitive because it won't cost him a fortune and it can be a skill that will help feed him for the rest of his life. Being self-employed with such a skill can even make you fairly wealthy. It is surprising how much a photographer can be paid for a single photo.

All of your gear should be purchased in a pawnshop, with

the best pawnshops being near military bases because young military people tend to have plenty of cash and act on impulse. They wander into the PX—a kind of department store with everything at discount prices—with pockets full of cash and buy nice cameras simply because they can. In many cases they use the camera for a short while and then lose interest, whereupon they often take the camera and lenses to the nearest pawnshop and sell them. Then they take the money and go buy something else, like a surfboard or stereo. If you aren't in or near a military town, pawnshops in cities like Las Vegas, Los Angeles, New York, Chicago, and other metropolises often have huge selections of camera gear for low prices. You might be amazed at how cheaply you can outfit yourself.

I make good money selling photos with magazine articles I write and to a stock photo agency (where my shots go for anywhere from $75 to several hundred dollars each), and I have only bought one camera new. The rest I purchased in pawnshops, and that includes many lenses, too. I would estimate that my photographs have helped, in one way or another, bring in somewhere around $50,000 in the last few years alone. And nearly all those photos were taken with cameras and lenses bought in pawnshops that were like new but which sold for nowhere near the new price.

If the pawnshop requires that you show some form of ID when buying something, always use an alternate ID. You can bet that they have security cameras watching you, so wear sunglasses and a hat to partially mask your facial features (but try not to look *too* shady and conspicuous; acting relaxed and natural is the key). Some heavy stubble, a beard, or a moustache that you can afford to shave off after leaving the shop are a good idea. Be aware that some cities now require pawnshops to have every customer leave a fingerprint before they purchase.

Once you are established, selling your photos can be everything from easy to infuriating. You have two options: sell them yourself directly to a buyer or sell them through a stock photo agency, which acts as a broker for your photos. I do both, and there are pros and cons on both sides of the equation.

The pros of selling your photos directly to buyers is that you have no middleman taking a cut and you eliminate another person who might know your name. The con is that you have to do all the marketing yourself, which includes getting every buyer's requirements and then trying to peddle your wares to him.

The benefit of using a stock photo agency is that all the marketing is taken care of for you. The con is that the agency takes a cut—often 50 percent of the price each photo sells for—and may know your name. You can, however, avoid the latter by operating under an assumed identity.

Photography is an excellent way to make a living *if* you master the skills needed to create a saleable, in-demand image and learn the business end. Remember, every time you see a photo in a newspaper, magazine, pamphlet, or book, someone was paid for it.

Personal Trainer

I know that someone is laughing about this line of work right now, but the truth is that good personal trainers make a decent living and are in demand. Like every other skill that brings home the bacon, the skills of a personal trainer are learned. But this is a type of work that you really need to be cut out for, because anyone who has been around a gym will know if you are the real deal or someone trying to bullshit his way through it.

For many years I was in this line of work, although in my case it sometimes fell as a collateral duty. I was good at it because I was in extremely good physical condition and knew how to teach and motivate my students. At the very least you must be in obviously good physical condition and know how to get others that way. This can be done by your doing two things, besides getting in shape: read every shred of information on physical training you can get your grubby mitts on—and there are plenty of magazines and books out there to help you do this—and start hanging around gyms and watching and listening to other trainers.

There are some former military personnel who are really riding this wave by running boot camp style physical training sessions for civilians who get a kick out of being yelled out and made to do push-ups in all sorts of weather at ungodly hours, like 0530. These guys have even been written up in major national magazines; that's how well they are doing (but do yourself a favor and *do not* get written up in any major magazine). This is a way of making good money without having to give a cut to a gym, because all your stuff can be done in a public park.

The drawback? You are going to have to advertise at first, which increases your exposure. This can be negated, however, by changing your name and appearance. It also puts you in touch with a lot of people who, if ever shown your photo by an investigator, may blurt out, "Sure I know him. He works down at the gym!"

"Marine Clean"

Here's a good one that takes little effort and training to get started. Feel free to use the name, too.

America has this view of Marines as being giant, ugly brutes brandishing weapons, kicking in doors, storming beacheads, and otherwise frightening people in the middle of the night. They also see Marines as having extremely clean barracks, with spit-shined decks and gleaming shower bulkheads. All of these perceptions are largely true, and you can create your own house-cleaning company that tells potential customers that a Marine or former Marine will come to their house and clean it so well that they will be most happy to pay you a decent fee for the service.

You are going to have to do some homework, like brush up on Marine jargon and know some Corps history, plus unit information and other things that will come in handy when someone asks you a question like, "What was your MOS?" (what was your job in the Corps?), "Where were you stationed?" and so on. Like an actor, you have to study your character and role and play the part to perfection.

House cleaning may be the perfect evasion profession because you only need a couple of houses a day to keep you in money, and you contact few people by doing so. The money is often under the table, too.

Independent Sea Life Harvester

If you have taken up digs near the seashore, for a small investment you can become a clammer (one who digs clams), sea urchin harvester (some sea urchins go for big bucks and are sold in Japan), wormer (one who harvests buckets full of marine worms on mud flats, which are sold to bait shops), lobsterman (one who traps lobsters in lobster pots and sells them to seafood dealers), crabber (you can run crab pots with a small boat, one that isn't nearly as expensive to buy and maintain as a lobster boat), oysterman (yes, you harvest oysters), fish-farm raider (done by sneaking into a commercial fish farm on scuba at night and helping yourself to the salmon or other fish being raised there; you can do this to lobster pots too), and so on. These types of jobs take only a little money to get into and a bit of research on how it is done.

Buying commercial fishing gear is best done through the want ads in the local newspaper and by shopping at yard, garage, and barn sales in seaside communities. The gear does not have to be pretty, only functional.

Depending on what type of commercial fishing work you will be doing, you will need everything from a flat-bottomed skiff (for worming and clamming on mud flats), an outboard motor, clam hods (elongate baskets), an oyster rake, open-slatted bushel baskets (for oystering, lobstering, and crabbing), closed-slatted bushel baskets (for worming), scuba gear (for sea urchins and for raiding lobster pots and fish farms at night), and a pickup truck for all these jobs.

This is a very good profession for the evader because payment is often made in cash, and you are your own boss.

Hull Inspector

If you are around large marinas with nice boats belonging to people who have far too much money, you can start a hull inspection business. This saves the boat owner the expense of having his boat hauled into dry dock. Scuba diving is easy to learn and become experienced in.

You must exercise caution when buying used scuba gear, since your life will be depending on it. Still, used gear is pretty easy to come by in most coastal areas, or you can buy it new.

You will need a tank (or dual tanks; aluminum or steel), regulator, depth gauge, dive watch, pressure gauge, wetsuit, fins, booties, mask, snorkel, weight belt and weights, buoyancy compensator ("BC"), and various gadgets and support gear like extra o-rings for your manifold and antifog drops for your mask. You can pick up an entire set of new scuba gear for under $1,500 easily.

You will also need a big underwater flashlight and a plastic slate with a waterproof "grease" pen attached, plus the usual stuff like business cards and a working knowledge of marine and diving terminology. The slate and pen are used to record things you find during your inspection that the boat owner will want to know about, like slime and barnacle build-up on the hull (increases drag when under way, which decreases fuel efficiency, and which can harm the hull's finish), dents, scratches, clogged vents, bent screws, and so on.

Like the sea life harvester, the hull inspector can ask for cash and act as his own boss, making it a discreet, low-profile job for the fugitive.

BARTERING

Bartering—trading one thing for another—has been going on for centuries and is still going strong today for the same reason it was going strong during young America's expansion westward, the Middle Ages, and the Roman Empire: everybody needs something sometime, and there have been and always will be those who know how to broker a deal. And bartering is conducted in every country, too, regardless of whether that country is a proverbial superpower or a rancid backwater dictatorship somewhere east of Egypt.

The traveling fugitive must become adapt at bartering if he wants to maintain some semblance of comfort and logistical stability. For instance, the time will come when, in some Third World country, you are low on greenbacks, which happen to be one of the most coveted currencies in the world today. It will no doubt be at just that moment that you find yourself in need of something: some line (rope) and a new screw (propellor) for your boat, a hot meal, medicinal drugs, shoes, or what have you. To the uninitiated, theft or a scam might be the only option that comes to mind, but to the professional fugitive, barter is the answer.

But where do you start? What is the *art of the deal*?

I've Got What You Want

Bartering is based upon mutual need; each party has something the other wants or needs, and a deal must be struck that is satisfactory to both parties, meaning that there will be no deal if one party thinks he is getting screwed. So, the answer is to find out what the other party needs and make him think you would really rather not give that item up, while at the same time not seeming overanxious to acquire what he has on offer. Yes, it is a game of wits and wills, and it is one you simply must become expert at. I have bartered literally all over the world and can attest to the fact that it is a fine way to get by, and sometimes get by pretty damn well. Here is an example of how it is done.

One day I found myself in East Africa where a gentleman had for sale a lovely carving of a reticulated giraffe which caught my eye. Did I need that damn giraffe? Of course not, but I *wanted* it, and that is what matters.

Step One: Appear to be merely a casual browser who has no clear agenda, that is, never stride right up to the shop, snatch what you like off the shelf, fondle and oggle it, then ask how much it costs. No, no, no! If you do this, the man will know you like it, want it, and will likely pay top dollar or trade something of great value for it. You must make him think that what he has is OK but isn't in any way important to you.

Step Two: Without picking the item up, get near it and make passing eye contact with it. If you pick it up you will have indicated to the man that you are at least somewhat interested in it, and the price will be set too high. If you make passing eye contact with it, he will detect this most minor display of interest and pick it up for you and hand it to you.

Step Three: Not too quickly—pause for a moment before accepting the thing—take the item and give it a cursory, half-interested inspection, then hand it back to the vendor and begin glancing at other items with the same halfhearted body language. Now, the man might not extend his hand to take it back. This is an old trick to get you to hang on to it longer, which with some people can make them more likely to take it off the guy's hands. If he tries this, set the item down and continue with the ruse.

Step Four: This is where your homework comes in. Before you showed up at the man's place of business, you should have determined what items or services the people in that area are likely to want or be in need of. (I once traded four Florida oranges for an entire woman, and she was worth every navel, I assure you.) Generally speaking, items that are frequently in demand in underdeveloped countries include medicine (aspirin, acetominophen, cough drops, isopropyl alcohol, iodine, etc.), fresh or canned fruit, canned meats, watches, knives, costume jewelry, reading glasses (the cheap kind you buy off a revolving stand in a drugstore), newspapers and magazines (in English; pornographic magazines, especially *Playboy*, are often extremely valuable), Nike sneakers, denim jeans

(Levis can bring a foreign vendor to his knees, I promise you, and the older, the better), pens, hats and T-shirts (particularly with American logos), Zippo lighters, and American cigarettes (Marlboros are tops).

This is how I do it. I put on a pair of well-worn Levis, Nikes, and a ball cap with a logo, and I put a couple of pens (Bic) in my shirt pocket along with a fresh pack of Marlboros and a Zippo (and I don't even smoke). On one wrist is a Timex watch and on the other is a cheap bracelet. On one finger is an equally cheap ring, but one with a big fake rock. In my hand I carry a canvas shopping bag that falls open when I set it down, revealing fruit, a *Playboy* still in the wrapper, a newspaper, and some other goodies. Now comes the coup de grace.

Step Five: Walk away.

The vendor is absolutely not going to let you get away that easy. He may even come running around his stall, gently grab you by the arm, and guide you back to his fine and upstanding establishment. Grudgingly oblige him. He will ask you what you are willing to give for the item he handed you and will tell you that, just for you, he will make a "special deal."

Watch his eyes. Are they glancing into your shopping bag, scoping out your jewelry, or scanning your jeans or other clothing? You have him.

Start low. When he hands you the item again, take it, look at it, and then set it right back down and indicate that you just aren't sure you want it. Before he has a chance to say anything, tell him that you might be willing to part with one of your treasures in trade for the item, but do *not* offer him anything he was looking at on your person or in your bag. Start with something small that he didn't show an interest in, which he will wave off.

Now, walk away again.

He'll stop you and will almost surely take another glance at what he wants. Catch his glance and immediately offer him that item. Now he knows that he has been caught and you can close the deal by turning the tables on him. Whatever he was looking at, hand it to him; if it was your hat, put it on his head and admire it. If it was the cigarettes, take one out, offer it to him, and stuff the pack into his hand or shirt pocket. Was his gaze wandering toward those precious Levis? You got it. Take them off then and there. (Of course you have shorts on underneath, just for this occasion.) Hand him the jeans.

Step Six: Extend your hand and ask if it is a deal. If you have played your part well and done your homework, you will walk away with the item you wanted, and each of you will feel you got the better of the other. It is fun and even exciting, and it works wherever you are.

The giraffe? I got it for a tuna sandwich with tomatoes and lettuce. Mmmm.

Bartering for services is a little different. You may be bartering for

- an item you want in return for a service you can provide
- a service you want in return for an item you have
- a service you want in return for a service you can provide

What might these services provided by someone be? Anything. From tuning up your engine or putting a cast on your broken arm to looking the other way when your boat doesn't quite have the right pollution-reduction equipment on the engine and not having the correct travel permit.

What services can you provide in return? You tell me. Use your imagination.

CHAPTER SEVEN

Personal Movement

> "You are conspicuous as hell."
>
> Paul Balor, *Manual of the Mercenary Soldier*

As a fugitive, you are a target. Whether traveling around North America or overseas, the idea is to become an elusive, "hardened" target that is impossible to pin down or predict. You must be:

- inaccessible
- unpredictable
- aware

To accomplish this, you must never follow a routine of any kind. Never do the same thing at the same time from day to day or week to week. Do not eat in the same restaurants, shop at the same stores, use the same route of travel, park in the same location, or do anything else that sets a pattern. In public, a hardened target thinks far ahead.

WALKING

Paul Balor, in his excellent book, *Manual of the Mercenary Soldier*, notes that even how you (a Westerner) walk is often different—noticeably different—than how locals in the Third World walk and move about. Westerners tend to be stiff and direct, getting to their destination in as straight a line as possible with eyes locked to the front and arms only swinging a bit. Balor notes that the "brisk, businesslike stride through town" of the Westerner is something that others can pick up on quickly, which can get you captured or dead just as quickly. Spend some time noticing how the indigenous people move about and then mimic them. Attention to detail is everything in this game.

You should frequently stop to "window-shop" and use the glass reflection to check for someone following you. The way you stop is important. Don't walk right up the window and stop without first glancing in. Instead, appear to just happen to glance in the window and then stop suddenly, as if something caught your eye. If you walk right up to the window, a tail will know he has been made and will break off, then signal another tail to pick you up.

Perhaps you have heard that another way to see if someone is following you is to glance in the passenger side side-view mirror of a car parked along the sidewalk. This is seldom possible, since such mirrors are almost always turned in toward the driver's seat and pointed at too low an angle to be of any use to you.

Never make eye contact with someone who you think is tailing you. This also alerts the tail and can make him nervous enough to try to take you down right there on the street. Public take-downs are ugly and cause you obvious problems. If you do make eye contact accidentally, don't panic and quickly look or turn away. Play it off as casual and even stroll over by the guy appearing to be interested in something or someone near him. If he's a pro and he thinks you made him, he will break off and call in another team to shadow you. If he is a novice, he will tighten up on you (follow you more closely).

Always have at least two viable escape routes, regardless of where you are at the time. Be watchful and ready. And *never hesitate* to make a break for it if you determine somebody is closing in on you.

Window reflections can be used to check behind you.

Walk away from all confrontations of happenstance. Never allow your pride or dignity to get in the way of your freedom. If you run into an asshole and decide to push the issue, you will make a scene and you could be arrested . . . and that's that. Walk away.

When on a sidewalk, walk facing the traffic so that you can see who and what is coming your way from the front. Walk on the inside of the sidewalk away from the street. The inside, although still risky due to possible snatch attempts from door-

At least six people in this photo taken in Vietnam are looking at the photographer, who is Caucasian. This should tell you something.

ways or alleys, is less so because most snatches and other attacks come from behind and on the outside so that the attacker can use the edge of the street for maneuver room.

DRIVING

Foreign driving laws and customs are often very different. For example, cars in heavy traffic in Egyptian cities like Cairo always part for pedestrians crossing the street. In Italy, pedestrians who raise one hand above their head and then walk straight into speeding traffic are not run down. In Germany, left turns against traffic are usually outlawed. Needless to say, if you were not aware of these common customs and laws you might plow down a local citizen, and what would *that* do for your low profile?

To lose a possible tail, slow to a stop at a yellow light when driving and then gun it through the intersection the moment the light turns red. Likewise, by taking two left or right turns when you suspect a tail, thus changing your direction by 180 degrees, you can pretty much verify or calm your suspicions. If the car following you also reverses directions, he is likely a tail. However, a good tail will recognize this trick and will break off, then call in another tail.

Always leave three-quarters of a car length between you and the car in front for quick pull-outs (a full car length is an invitation for someone to think you are allowing them to cut in front of you).

Common-sense security measures are a given. Your car must always have at least half a tank of gas and be in prime mechanical condition. When you park, back in. Scan the area first when returning to your car. A remote keyless entry device is terrific so that you need not fumble with keys. As you approach the vehicle, be alert and look around. Try to approach the vehicle from an open area to help avoid surprises hiding nearby. If other people are around, all the better, but watch out for the trick of two or three seemingly innocent people walking past you and then jumping you from behind. Of course, checking the backseat for an uninvited guest is mandatory. Also, if the car doesn't start right away, immediately suspect that it has been tampered with and get out of there.

Roadblocks

Roadblocks/security checkpoints call for luck, bluff, and preparation. Oftentimes if you act like you know what you are doing, you will get through. However, sometimes the mind-set of the roadblockers is such that you are going to have to bribe them or prove you are authorized to pass with bogus paperwork. These can be very difficult situations. I was recently speaking with my friend, Mark, who told me about the time he was in Angola years ago during troubled times. He came upon a roadblock and simply drove right through it, waving and smiling as he went. The soldiers just waved and smiled back. He decided to do this because the day before he saw another white man do the same thing. Hey, if it worked once . . .

AIRPORTS

As an international fugitive running from whoever or whatever, you need to think carefully about air travel. Sometimes it is a good option; other times it is a terrible choice that must be avoided.

Airports are one of the most dangerous places for the fugitive since they are crawling with people whose job it may be to catch you. They are also bad for running into old acquaintances who recognize you and shout out your name with big, stupid grins on their faces.

Arrive at the airport just in time for your flight. The goal is to spend as little time in the airport as possible. Buy your ticket through a travel agent who can issue boarding passes at the time the ticket is purchased (further reducing time in the terminal). Electronic ticketing is best of all because it allows you to go straight to the gate if you have no baggage to check.

Know what other flights are leaving for your destination about the same time as your flight. When you check in, ask if any seats are available on the other flight and make the switch if possible.

Don't check luggage in; use carry-ons instead to reduce time at the airport on the other end and eliminate the chances of someone scanning the checked luggage for your name (luggage should never have your name on it anyway; mark it with a special mark known only to you, such as a small piece of tape). Airlines are getting more restrictive on carry-ons, however. Today you may be allowed only one, and it might have to fit through a narrow opening in front of the X-ray machine. Call ahead and plan accordingly.

Don't drink alcohol at the airport; you need to be very

sharp at all times. Besides, drinking at an airport bar can lead to unnecessary and unwanted conversations with lonely, chatty, or overly curious travelers or, worse, trolling security types.

Not all airlines are created equal insofar as searches, security, documentation, reliability, prices, comfort, safety, and flights available. And whereas it is true that airport and airline security in America is getting tighter, overseas it may be another story. It all depends on where you are. The Middle East, Greece, Africa (especially Lagos, Nigeria), Asia, parts of the Orient, and much of Central and South America are very lax when it comes to checking identification.

International airports, however, can present all manner of problems, such as the time my friend David had a little weapons problem when coming into Zambia via the airport at the capital of Lusaka. The Zambian chief of police at the time was one Colonel Umbulu, whose appearance is reminiscent of Idi Amin, the former Ugandan dictator, thug, and cannibal. David managed to bribe his way out of an extended stay in a rather nasty prison cell, or perhaps a bullet between the eyes, by arranging for 100 pounds of dried Cape buffalo meat to be delivered to the colonel. (David, an experienced Africa hand, would also advise you to have many smaller bribes available—cigarettes, liquor, cash, newspapers, magazines—to grease soldiers and policemen at the many roadblocks you will no doubt run in to when traveling overland. These same rules apply to bus and train travel, too.)

HOTELS

Stay in hotel rooms between the second and fourth floor for ease of escape and security. These floors have flights of stairs that are easy and quick to get down, which makes it more handy for sudden escapes. Ground floors are the most easily accessed and have much more foot traffic, which can inadvertently make you lower your guard because you may become used to the sound of others walking nearby.

Select your room wisely. Get one that is easily escaped from.

Change your reserved room upon check-in at the hotel, if possible. If you are not expecting any visitors, tip the desk clerk well and tell him that you are not registered there. Let him know that another tip will come his way when you leave if he agrees to this. Have him delete your name entirely from the registry if possible, but make sure he puts someone else's name down for your room so that another guest isn't assigned the room by the on-coming clerk.

Do not answer a page by the hotel. You are always out.

Meet contacts away from the hotel at a place of your choosing. The lobby is the first place someone who is looking for you will stake out. In fact, never even enter or leave the premises via the lobby; come in and go out a back or side entrance, and first surveil the area outside the door you will use to see if anyone is waiting for you. Know where all exits are in the hotel.

Do not eat or drink in hotel establishments or use any of the other hotel facilities like laundry service, the pool, and so on. Using such facilities puts you in view of others more often, and somebody may remember the gringo when questioned later. The fewer people who see you, the better.

Keep your bags 100 percent packed at all times. You need to always be ready to leave the hotel quickly. Never unpack.

The drapes in your room should remained closed. Frequently look out from the side without moving them. When leaving the room, leave a hair across the upper jam by wetting it with saliva where it touches the door and the jam, or place a small piece of string on top of the door. If it is gone when you return, you know someone has been in the room. Of course, it could have been the maid.

Never forget that you are vulnerable when in a hotel room. I always lock the doorknob, use the dead bolt, and secure the chain. I then prop a chair under the doorknob and wedge a small screw between the jam and door, which makes it more difficult to open forcefully from the outside. I often carry a rappeling rope with me and tie it to the bed frame. If someone is trying to break in via the door and I have to leave by some other means, I can body rappel down the rope to safety. (For those of you who may be unfamiliar with this technique, tie intermittent knots in the rope to facilitate climbing down it.)

Be aware of security cameras.

DINING OUT

Sit in the back of a restaurant near the rear exit. Face the main entrance while keeping an eye on the rear exit at the same time. Check the rest rooms (all of them) for window exits.

If possible, get a table that places a stanchion or other obstruction between you and the main entrance. This blocks your opponent's immediate clear view of you but allows you to glance around it frequently. The split second you gain here might be the split second that saves your life.

Never dine *al fresco*, that is, outside on a patio or sidewalk cafe. This allows too many avenues of approach and also makes it much easier for someone to spot you as they walk by or from a distance. You might even be caught by a security camera that surveils the entire area.

OTHER GENERAL TRAVEL RULES

You must be physically and mentally prepared to silently kill or incapacitate an enemy at a moment's notice. You must, therefore, always be armed in some way and be ready and able to use improvised weapons, since there will be times when a gun or knife is unavailable or ill-advised, such as on an airplane. (Actually, it is ill-advised to incapacitate *anyone* on an airplane for *any* reason using *any* method! Where are you going to go after doing so? Face it—if you are confronted on an airplane, you must either bluff your way out of it or the jig is up.) There is a way around this, and that is by carrying a knife or other weapon that is nonferrous—has no iron in it. Most metal detectors search for iron, making something like a 100 percent titanium knife undetectable to the metal detector. A friend of mine recently tested this theory with a Bowie knife made of titanium. He sailed right through security with the knife tied around his neck and concealed under his shirt.

Improvised weapons are always at hand. A pen or pencil, comb, belt, rolled-up magazine, paper clip, glass, spoon, key, and many other everyday objects can quickly and easily be used as a weapon. Think ahead and use your head. I suggest you peruse the Paladin catalog for additional info on improvised weapons, as going into depth on them here is beyond the scope of this book.

Pay in cash so that you leave no paper trails of traveler's checks, personal checks, credit cards, or debit cards in your wake. It is very easy to lay such a trail, and a professional tracking you will have no problem picking up your paper trail if you leave one.

One way to throw off an investigator is to arrange to have a trusted associate scatter your credit cards in various spots such as college campuses and crime-ridden neighborhoods. With any luck, the finder will go on a spending spree that will leave a false paper trail for any nosy investigator. A good one will catch on to this old ruse quickly, but it just may buy you a crucial day or two to make your getaway. (By the way, never drop the whole wallet; the chances are fair that Mr. Honest John will return it to you, nullifying the entire operation.)

Carry a good "alternate" ID card with you. Many are the fugitives who have tricked those hunting for them by whipping out a convincing alternate ID. You might be surprised how this simple ruse can work so well in an emergency. This can be as simple as a stack of business cards cranked out at a local print shop or as elaborate as a phony passport with supporting documentation acquired through underground channels.

Dress like and follow the customs and habits of the locals if at all possible. If you are white and traveling in East Africa, you won't get away with dressing like a traditional Masai. However, you *can* get away with dressing like a tourist on a photo safari. Common sense always applies.

Use the local currency exclusively. Never flash dollar bills. Exchange your U.S. dollars for the local currency at the airport on your way out.

Do not display any tattoos, scars, or other distinguishing marks. Laser surgery works wonders on tattoos and even some scars. Moles and other distinguishing marks can and must be removed. A box of hair coloring is available at any drug or discount store for less than $10. Buy some nonprescription colored contact lenses in a shade other than your

Sunglasses. Uniforms. Radios. Could these guys be looking for you?

natural eye color. Nonprescription glasses also help change your normal appearance.

Do not use ATMs without a disguise on. Just about anybody who really wants to can get access to the videotape from the security cameras attached to each machine. Likewise, all establishments with security cameras are to be avoided. The best idea is to always be somewhat disguised—never appear in public as the real you.

If you are good enough at all this, those who want you will never find you. For example, I recently received a note from a very famous fugitive—perhaps the most famous successful fugitive ever—who hasn't been seen in North America since the 1970s. A great many people are still looking for this daring individual, but I doubt they will ever catch up with him. By sheer coincidence, I am very familiar with the country and region he is living in—or at least where his note was postmarked from—and must say that I, too, would have seriously considered this area if I were in his sandals.

Conclusion

> "And the three men I admire most,
> the Father, Son, and the Holy Ghost,
> they caught the last train for the coast . . ."
>
> Don McLean
> *American Pie*

It happens every day to everyday people like you and me. You have a sudden, unexpected spell of very bad judgment. You piss someone off and they set you up to take the fall for a crime you did not commit. You witness something that you wish to God you had never witnessed. Your personal finances go south and you have to start over somewhere else and as someone new. Your legal bills are a mile high and you can't pay them, so your friendly lawyer puts liens on everything you own. You get involved with the wrong crowd and now you are being sought by people who will seriously hurt or kill you if they find you. You are in an abusive relationship and have to get out of it to survive. Or perhaps you just want to forget your past and begin anew. Couldn't happen to you? Such things only happen to the other guy? That's what he is thinking; to him, you *are* the other guy.

The fugitive's world is much more than alleged

killers and certified losers like Eric Rudolph, the Four Corners fugitives, and Andrew Cunanan. It is about the real world where bad things happen to people who never expected to become the victim of laughing chance. How you handle the situation will dictate precisely what your future holds, be it a new life as a photographer in Scotland, a jail cell, or a grave.

Preparation and immediate action are the key. If you take the time and spend the money and energy to prepare yourself for life on the lam, you can make it. And if Lady Luck smiles on you from time to time, you might end up with a life that is better than your last one.

Running, hiding, and surviving might seem like a daunting task, and a long-lasting one at that, but consider all the successful fugitives who are out there and free despite massive manhunts conducted by thousands of people, such as the searches for the Four Corners fugitives and suspected bomber Eric Rudolph. Sometimes those hunting for you can actually assist in your evasion, as was the case with both of these searches. If I were a fugitive again, I would dearly love to see hundreds of inept searchers tramping through the countryside, destroying all evidence of my passing, and posing for the television cameras. I would love to see chiefs of police bragging to media types about how they have me surrounded and that it wouldn't be long now. Better yet, it would be like Christmas if they set the area on fire in order to smoke me out, as the bumbling chief of police of Bluff, Utah, did in his ridiculous and vane attempts to capture the Four Corners killers. Fire can destroy evidence very well indeed. And perhaps the best help one can get from searchers is when the feds are called in. When they call out the National Guard and are convinced that they have you very nearly cornered, as was the case with Rudolph and the Four Corners fugitives, you are practically assured of your continued freedom.

In closing, remember this: scared little bunnies are able to survive in the big, bad woods because they know that their only hope is to remain out of sight, sound, and smell of the hawk.

You do the same.